# The
# ...
# to
# Online Services

# The Woman's Guide to Online Services

Judith A. Broadhurst

## McGraw-Hill, Inc.

New York  San Francisco  Washington, D.C.  Auckland  Bogotá
Caracas  Lisbon  London  Madrid  Mexico City  Milan
Montreal  New Delhi  San Juan  Singapore
Sydney  Tokyo  Toronto

©1995 by **Judith A. Broadhurst.**
Published by McGraw-Hill, Inc.

Printed in the United States of America. All rights reserved. The publisher takes no
responsibility for the use of any materials or methods described in this book, nor for the
products thereof.

pbk   1 2 3 4 5 6 7 8 9 DOC/DOC 9 0 0 9 8 7 6 5

**Library of Congress Cataloging-in-Publication Data**
Broadhurst, Judith A.
        The woman's guide to online services / by Judith A. Broadhurst.
            p.    cm.
        Includes index.
        ISBN 0-07-024168-6 (pbk.)
        1. Videotext systems.  2. Internet (Computer network)      I. Title
QA76.55.B76      1995                                              95-17373
025.04—dc20                                                       CIP

Editorial team: Brad Schepp, Acquisitions Editor
                Kellie Hagan, Book Editor
                Robert E. Ostrander, Executive Editor
Production team: Katherine G. Brown, Director
                Rose McFarland, Desktop Operator
                Linda L. King, Proofreading
                Nancy K. Mickley, Proofreading
                Jodi L. Tyler, Indexer
Design team: Jaclyn J. Boone, Designer                           0241686
                Katherine Lukaszewicz, Associate Designer         BR1

# Contents

## Appendices

# Acknowledgments

My first thanks must go to the people who made this book possible: Carol Nelson, who suggested this book to McGraw-Hill; Brad Schepp, who chose me to write it; Brad, Judy Heim, and Kellie Hagan, who supported my desire to depart from the usual computer book formula; and Judy Kessler, Cynthia Borg, and Barbara Hendra, who enthusiastically took on promotion. No less gratitude goes to Stacey Spurlock, Linda Poon, Jackie Boone, Kathy Gilligan, and all the rest of the staff at McGraw-Hill for treating my requests and idiosyncrasies kindly and coming through, time after time.

As corny as I always thought it sounded, I now understand why first-time authors, particularly, feel so grateful to those who guided and helped them at turning points in their lives. So heartfelt thanks to my late stepfather, Philip Sieg, who first taught me the value of well-crafted prose, and to my mother, Nina Kinney Sieg, who finally understood, stopped suggesting I get a "real job," and said instead, "You're a writer; that's what you do." Some of the following people might not even realize that their support, both personal and professional, made a difference, but it surely did: Jim Cameron of CompuServe's Journalism Forum; editors Rick Chatenever, David Crossley, and Sue Hawthorne; Carola Draxler; Doug Eads; my sister, Joan Akey Grace; Angela Gunn; Susan Heinlein; Elizabeth Jay; Georgene Lockwood; Mark Loundy; Teresa Mears; Ellen Mitchell; Alan Rowberg; Charles Shaw; Michael Smiley; Doris White; and Louise Williams.

Just as important to this book were those who helped with it at crucial junctures: Paul Aiken and Ed McCoyd of the Author's Guild, Paul Edwards of CompuServe's Working from Home Forum, Phil Mattera of the National Writers' Union, Joe Motyka and Tom Potts of U.S. Robotics, Alec Saunders and Jodi DeLeon of Microsoft, Leah Reichman, and Esther Schindler and Sandra Donnelly of ZiffNet. Blessings upon my computer mentors and troubleshooters who, thanks to modems, are in California, New York, and London: Jared Sherman, Roman Iwaschkin, Steve Kress, and Peter Wilde.

Very special thanks to everyone who agreed to an interview for this book, including many whose names don't appear, and to the subscribers to my newsletter, *Freelance Success*, who stuck with me and even cheered me on when completing the book took precedence over getting the newsletter to them on time.

Without review copies of software and loaned hardware, this book would have been far less comprehensive and a lot harder to do, so I sincerely thank the many companies who provided the following:

- A Macintosh Centris 650 computer system and software for eWorld, from Apple Computer

- BIBL bibliographic software from GMUtant Software

- Hijack Graphics Suite production and conversion software from Inset

- Internet in a Box software from Spry

- Journalist customized news software for CompuServe and Prodigy from PED Software Corporation

- A NEC MultiSync XE15 monitor, monitor lens, and AudioTower speaker system from NEC Technologies

- Netcruiser software from Netcom

- Netscape software from Netscape Communications

- Norton AntiVirus software from Symantec

➤ Sound Blaster 16 sound card and speakers by Yamaha from Creative Labs

➤ Sportster 28.8Kbps v .34 modem from U.S. Robotics, Inc.

➤ Toshiba color laptop computer loaned by Toshiba American Information Systems

➤ Western Digital IDE hard disk drive

➤ WinComm communications and WinFaxPro fax software from Delrina

Thanks also to the people who provided or authorized screen shots. And particular thanks for all the help from the staff and sysops of:

➤ America Online ( The logo on the back cover of this book is copyrighted by America Online; used with permission.)

➤ CompuServe

➤ Delphi

➤ Echo

➤ eWorld

➤ GEnie

➤ Interchange Online Network

➤ Microsoft Network

➤ Netcom

➤ Lexis-Nexis

➤ The Pipeline Network

➤ Prodigy

➤ The WELL

➤ Women's Wire

# Preface

Your first reaction to the idea of a book called *The Woman's Guide to Online Services* might be the same that mine was: With so many online guides already, why do women need a special one? Let me assure you right off that this is not a dumbed-down book nor one solely about women's sections online, lipstick and lace, and home and hearth disguised as "women's issues." Nor is it a technobabble treatise on technology or another list of lists.

The key word in the term *computer network* is *network*, not *computer*. So this is a book about how to work the networks, rather than how to make the networks work. It's also about ways to use online services that are relevant to real-life, day-to-day concerns, and how to make your life richer, easier, and more fun. Computers are designed to save you time, so it's about how to do that too. It also answers the three basic questions that the hundreds of women I've interviewed for this book and for articles about online services for *Executive Female*, *Glamour*, *Mobile Office*, *Online Access*, *Self*, *Working Woman*, and other magazines have told me they want to know:

➤ What's there, once I get online?

➤ What can I do with online services that's truly worthwhile?

➤ What are the fastest, easiest, most fun ways to do those things?

Throughout this book, both women and men will tell you, in their words, how they've used online services to make significant, meaningful differences in their personal, family, community, and work lives. From their stories, you'll get a feel for what it's like online from the human side and see that, as Lisa Kimball of The Meta Network says, "The relationships online are much more where the power is than in the information." Then each chapter shows you how to find and use the resources related to that particular topic. Directories at the end of most chapters list related sections on America Online, CompuServe, Delphi, eWorld, GEnie, Interchange, the Internet, Microsoft Network, Prodigy, Women's Wire, and the World Wide Web, plus smaller networks such as Echo and the WELL. These will give you more specific information about more services than any other online guide, all from a woman's perspective, but all useful to men too.

But I didn't write this book just to convince you that online services have more to offer than casual conversation and information, although that's certainly one thing I hope it accomplishes. I wrote it because the ratio of men to women online worries me. A lot. As this book goes to press, in early 1995, the ratio of men to women with their own accounts on CompuServe is 9:1. The rough estimate of the population of women on the Internet is 10 percent, and most of those women are in the high-tech industry or academia, including students with free accounts. Prodigy, the most family-oriented service, claims that almost 40 percent of its members are women, but industry experts doubt that claim because Prodigy always waffles when you ask how they came up with their number. Even giving them the benefit of the doubt, they apparently count everyone in a household who has access, which assumes that women use the service too. Maybe, maybe not. Women's Wire, at 90 percent, and Echo, with 38 percent, are the two services covered in this book that are most likely to really have the high proportion of women that they claim, because they're owned by women who have intentionally created places where women feel welcome and comfortable.

*The relationships online are much more
where the power is than in the information.*

This disparity in the ratio of men and women online worries me because I'm old enough to remember when women fought so hard and so long to gain what often seemed like so little. Now I see us losing ground because of sensationalized media hype, stereotypes, and perhaps the plain stubbornness and short-sightedness of women themselves. I'm absolutely, thoroughly convinced that the continued reluctance or resistance that so many women clearly have about going online is already to our detriment, and is likely to have more serious, negative ramifications in the long term. It's not merely a matter of career consequences, or even the growing distance between the haves and the have-nots. Ultimately, it comes down to power. Michael Korda warned us 20 years ago when he wrote *Power: How to Get It, How to Use It*: "The person who controls the computer is thus in a singular position of power." He then went on to describe how that person gains power over not just information, but people. Korda's book was published in 1975, just three years after the first public demonstration of what we now call the Internet and four years before the oldest commercial service, CompuServe, existed. Today, it's even more important that we heed his words.

Because I'm a journalist by both profession and nature and, like many other journalists, wanted to be one so I could do something modest like change the world, I always remembered what Korda said about the power that comes with controlling information. So Korda's book has remained on my bookshelf all these years, even when I recycled many others at used bookstores. Since going online myself in 1990, I've often wondered why computers have come to be considered such "a male thing" and why online services are such a male domain. But I never made the connection until I read *Neuromancer*, the 1984 futuristic fantasy by William Gibson, who coined the term *cyberspace*. Gibson's hero thought of himself as a "computer cowboy," and used his computer to catapult into the nebulous realm of cyberspace to fight the bad guys. He called the process of using his computer to transport himself into cyberspace "jacking in." The moment I read those words, I understood. If you'll pardon the allusion, men get off on the power that truly skillful use of all the computer's capabilities gives them. Jacking into cyberspace—a term now commonly used on the Internet—makes them feel powerful and adventurous, like cowboys on the new frontier. It's no surprise, then, that when a cadre of high-powered cyberspace gurus (all men) created an organization in 1990 to defend the independence and traditions of the Internet, they named it the Electronic Frontier Foundation.

Cyberspace is still a frontier, which is good. That means there's still time for women to be pioneers too, and help build this new territory. This *is* the "global village" that media critic Marshall McLuhan wrote about, more so than anything we've ever experienced before. That might sound grandiose or fanatical, but what happens online has already had an impact on the real world and will affect what goes on hereafter in many ways in our homes, schools, communities, and countries. If you doubt that, I hope this book will help you see some of the possibilities and what they mean for you and your family, for better or for worse.

Because this cyberspace world is going to affect us all, whether we're online or not, it behooves us to help guide its development into the kind of place we want to be involved in as we grow old, and for our children to be a part of. You might not care about having influence or power, per se—no one but Donald Trump ever admits they do—but you probably *do* care about your education, job, the environment, and world peace, right? All of these will be profoundly affected in the years to come by people communicating and exchanging information online. If you want to have some say in what happens, you need to be where it's happening and connected to the people making it happen.

If, as some studies say, women are more verbal than men and communicate better, then it's our kind of world, because communication is the essence of what being online is all about. Until more women stop thinking of computers either as mere machines or as being too complicated, and start using them as the powerful communication tools they are, we're relegating ourselves to the techno-era equivalent of the typing pool. We're also missing an important opportunity to help shape the future. Men have repeatedly helped me learn my way around online, so I'm not saying that they're in any way hindering us. We're the only ones holding ourselves and each other back. It's up to us to put aside our own skepticism, wariness, and "yes, but . . ." attitudes and move forward.

That's why this book is a *woman's* guide to online services, in the singular. Because the way to keep cyberspace from being the latest good ol' boys' network is for more women to get online and get active, not just lurk or observe, until we all truly understand it, feel good about it, and feel like we belong there. And the way to do that is one woman at a time.

# Introduction

## Why there's a male mystique about modems

Warning: The following statements are sure to offend someone.

Because I've admittedly been a bit of a fanatic about getting other women online, I've interviewed or talked with a few hundred women about why they are or aren't online and tried to keep up on relevant research. People continually ask what I've learned and how it all fits together, so here's how I see it. Just bear in mind that what follows are mainly my impressions and opinions, other than direct quotations from other people. Informed opinions, yes, but based as much on anecdotal and empirical evidence as on results of actual research by social scientists. Nonetheless, the patterns and progression are pretty clear.

A businesswoman I once interviewed told me that she had concluded that online services were useless for anything she might want to do, as far as she could tell. She explained that she had tried CompuServe a few years back and found it too much of a hassle to learn to navigate (this was before the automated software they now offer), and that there was nothing there that she couldn't get more easily elsewhere. The weather report, for instance, is easier to get with a local phone call or from The Weather Channel on cable TV, she said. Or airline reservations. She'd rather just call her travel agent. There's merit to that opinion, I said, but what about the forums and research databases? Oh, she hadn't tried those, she replied, because they cost extra. I was dumbfounded. That's like going to the circus, but never going beyond the parking lot, and judging what a circus is like even though you've never actually seen the circus itself.

Another woman told me that she didn't want to get online because "The idea of communicating with someone through a computer sounds so sterile and mechanical." At the time, this woman was the president of the New York City chapter of a very technically oriented association for women in the computer industry, so it certainly wasn't fear or computerphobia that was really at the root of her resistance. But feeling fearful, intimidated, or unsure of themselves is the deep-down reason some women resist going online. According to a 1994 survey for MCI by The Gallup Organization (see Fig. I-1), 27 percent of men and 39 percent of women see themselves as cyberphobic. According to a February 1995 estimate from Jupiter Communications, only 20 percent of the people online are women, and they expect that to increase to only 38 percent

I-1 *From MCI's Cyberphobia survey, conducted by the Gallup Organization in 1994.*

by 1998. So that 12 percent difference between men and women in MCI's cyberphobia study isn't enough to account for the 60 percent difference between men and women currently online.

The reasons behind those figures are more complicated than what people usually assume, as things often are. One of the obstacles is how women think of computers in general. Research results indicate that men see computers as technological toys, therefore fun, but women see them as mere machines, thus associated with work. Too many women still don't think of them as communication tools, like telephones. As Lee Sproull, a sociologist and researcher at Boston University's School of Management, says, "Women tend to think of computers as productivity tools, whether it be word processing or spreadsheets or database management, rather than something that connects them to other people."

# It all starts with tea parties and erector sets

Part of that mindset goes back to childhood, says Mary Frances Stuck, a sociologist at the State University of New York at Oswego. "Boys tear things apart and put them together again and use them in different ways than what they were intended for, but girls think only tea goes with the tea set, not chemicals."

Of course you've probably already heard a couple of the other factors mentioned many times: Until recently, computer games have been designed by big boys for little boys, and the themes are violence, killing, and far-out fantasies. Most of them are still shoot-em and shred-em products, and this turns girls off. And just as it's the man in the household who's most likely to buy a computer, even though it's more likely the woman who uses it to manage the family budget, it's the boys whom parents most often encourage to use computers.

"Parents are more likely to buy a computer for boys, and more likely to send their male kids to computer camps," says Chuck Huff, an associate professor of psychology at St. Olaf College, in Northfield, Minnesota. "These kinds of biases are deeply ingrained in our society, and they're going to be hard to get out just because we're going to be nice."

There's also a significant shift that occurs in girls' experiences with computers at school around junior-high age, and affects them thereafter. Up until that time, research shows that girls are just as interested in computers as boys are, and do just as well at learning to use them. But once the hormones start flowing, the dynamics change. When a girl in a school computer lab has a question about how to do something on the computer, a boy often takes over and does it for her, in an effort to impress her. To assure him that she appreciates his help and is duly impressed by his superior ability, she just sweetly says "Thank you." Thus she doesn't continue to learn and gain confidence at a comparable pace, as she had before. Moreover, boys often tend to stake their claims to the few computers available and take up most of the time in the computer labs. Teachers too often let them, which leaves some of the girls sitting on the sidelines, waiting for their turns at the keyboards.

"Young girls seem to think it's fine for girls to become computer people," says Huff, "but young boys don't." Most girls get the message and find other interests.

# Women bring what they've learned to the workplace

Eventually, what began when they were children carries over for women as grown-ups. "As adults, males tend to look at computers as things to be played with and explored," says Stuck, "whereas women want to know precisely what computers will do and what keys to push to get them to do that."

In the workplace, a different male-female pattern comes into play. Perhaps because so many women came up through the ranks and had to do much of their own clerical work, they continue to do more of it than men, and continue to think that's the purpose of computers: to crunch numbers, compile data, and write words.

And even though computers are now commonplace in offices for everybody from the receptionist to the CEO, men and women use them differently. Men who previously claimed they couldn't type have taken to computers like preachers to the pulpit, but they use them more to analyze data, collect online information about competitors, and wheel and deal by e-mail. Women, on the other hand, repeatedly tell me that they feel guilty if they use their computers at work for anything other than what's obviously "work." Unless their bosses specifically instruct them to do so or they work in the research or MIS department, they seldom see browsing online services for information that could help them do their jobs better or help the company compete better as "real work." Instead, many of them think of it as the equivalent of chatting with a friend by phone on company time.

In a conversation about this observation, Susan Herring, a linguist at the University of Texas at Arlington who does research on how men and women communicate online, told me I'd be reinforcing a stereotype if I said this. Maybe so. But stereotypes are rooted in reality. So many women have told me variations of the "it's not real work" story that it seems to be a common belief, and warrants actual research in real-world situations. But real-world research isn't popular, nor is it easy. With the notable exception of a study by Lee Sproull and Sarah Kiesler about the effects of e-mail in the corporate culture, the studies about women and computers or online communication have been done on the Internet or in controlled situations such as classrooms or campuses, all of which are skewed views, and not sufficiently relevant to real life.

One respected authority told me that academicians simply can't afford to do research on the commercial networks. Balderdash. They can get grants, and if freelance writers, who aren't known for making big money, fork out the fees for commercial services because they feel they're important to their work, this is a weak excuse for sticking to the easier path and perpetuating a distorted picture of what's happening.

Now that the commercial nets and the Internet are converging, it might be too late to get any meaningful comparison of the cultures. Maybe it doesn't matter anyway, because my major concern is that more women get online so we're not left behind.

# The seven biggest factors that keep women offline

In addition to all the historical factors, and probably more important than any of them, there are at least seven more reasons that more women haven't yet gone online. Virtually all involve misperceptions and stereotypes:

## Lack of time

Women lament that they never have time to be online, either at work or at home. But, particularly at work, that's partly due to how they handle their time. Even though they now have the tools to do the tasks themselves, Stuck says that men still delegate more than women. They're more likely than women to use their time for things other than what's basically clerical work, because clerical work is still seen as women's work when push comes to shove.

This time problem carries over into home and leisure lives too, and plays a part in why more women aren't online. Lack of time, in fact, is one of the major reasons women cite as why they aren't online, and it's a valid one. "Somebody's still doing the dishes and putting the kids to bed and doing the errands during the evening hours when most people are online," says Janet Attard, who leads the small business forums on both GEnie and America Online. "That person is usually still the woman, and she just doesn't have the time to spend online." Rather ironic, isn't it, when computers are intended as time-saving devices? "Getting over the learning curve used to be the stumbling block," Attard adds, "but now, once you get beyond getting the modem installed and the computer to talk to the modem, it's quite easy. You see a picture, and just point to it and click."

The assumption has been that women were simply techno-Neanderthals, inept with technology, or computerphobic. Thus the major credit for more women going online since the boom of 1994 has gone to those point-and-click graphical user interfaces. They have been a major influence, certainly. But as much of the reason for women's reluctance to boot up and log on is tied to the time issue as it is to the technology, so just the fact that newer programs take less time to install and learn makes a difference.

## Attitudes about computers

A lot of us simply don't care about all this folderol. We don't find fiddling with CONFIG.SYS files or IRQ switches fun at all. As Paula Span put it in an on-the-mark essay for the *Washington Post Magazine*, "Women treat computers like reliable station wagons: learn how to make them take you where you want to go, and as long as they're functioning properly, who cares about pistons and horse-power? Computers are useful but unexciting. When something goes wrong, you call a mechanic. Whereas guys, even those who never learned how to change an oil filter, are enamored of computers, want to play with them, upgrade them, fix them when they falter, and compare theirs with the other guys'."

## The geek chic stereotype

"Before I was online," says Esther Gwinnell, a psychiatrist in Portland, Oregon, "I'd imagined it as mostly people who build computers, with pocket protectors and many pens. Boy, was I wrong!" The techies are there, of course, but all online services are so mainstream now that this is simply an outdated, off-base impression.

## What we want isn't what's there

According to a report in *Inside Media* magazine, when someone at a public presentation on online communities asked Steve Case, CEO of America Online, what else it would take to get more women online, aside from the appeal of community with other people, his one-word answer was: shopping. Tell that to any woman and watch her bristle. Sure, we want online shopping if it saves us time and makes our lives easier, and while you're at it, give us photos, not just text lines like "Ralph Lauren towels, Style 487, $7.99" or "red dress, $59." But we're long-since tired of being classified as consumers, thank you. What women want to see, to be convinced, are ways to use online services that apply to their real-world lives—not just chat and further information overload—and content online that's interesting without being insulting. Most of us are interested in the same things men are, so quit touting the recipe databases, please.

# Misdirected marketing

Says Judy Heim, a practical woman who writes for *PC World*, "Online services have marketed themselves on the basis of gee-whiz features like online banking, electronic encyclopedias, and big databases, but the real virtue of online services is the forums. That's where all the good information is to be found, and women are sometimes slow to catch on to this. Part of the reason there aren't more women online may be that women are shrewder consumers than marketers think they are. Their reactions are often something like, 'You're telling me I can dial up an online service and pay an extra $15 a month to balance my checkbook? Yeah, right.'"

# A matter of economic priorities

For many women, especially single mothers, working or not, computers and online services are costly luxuries. Even those who can afford them have other priorities for their money. "I think women are less eager to spend money on hobbies," says Heim. "We consider it self-indulgent. We feel guilty, thinking we really should be spending the money on our kids or other loved ones."

# Scare stories from the media

It's understandable that people have bought into this last stereotype, and it's probably the single biggest reason why more women are not online. It's the mythology the media have fed us that online services are populated by sexual predators; rude, crude, and lewd men; and social misfits. The blame—and I have no qualms in calling it blame—lies squarely with the mainstream media who have been nothing short of grossly irresponsible in running exaggerated, distorted, misrepresented stories that have flat scared many women away. A search for *women* and *online services* on any magazine or newspaper database or in any of the archives of major online publications invariably turns up mostly a list of predictable stories about sexual harassment, pedophiles, stalkers, cybersex, online romances gone wrong, and the theoretical risks of using credit cards for online shopping. These make for juicy copy on a slow news day and build readership, which increases subscriptions and boosts ad revenue. Fine. But that's certainly no excuse for pandering to prurient tastes.

Well, ladies, you've been misled. Boondoggled and bamboozled. Some of the weirdo stuff that's reported exists or has happened, but rarely, which is precisely what makes it such juicy copy. Much of it, however, is pure sensationalism. Can you imagine putting 20 million people together anyplace else on earth where some of this wouldn't happen, and far more often? This yellow journalism has kept thousands of women from even trying online services. So I'll address some of the rumors, myths, and hyperbole in chapter 1, which is about what it's really like online, contrary to what you might have read and heard, and why it's worth being there.

# CHAPTER 1

# What it's really like online

**A**ll the brouhaha about online sex and safety and the peculiarities of the "online culture" have given people who aren't online the impression that cyberspace is as weird and other-worldly as the name sounds, rather daunting, or just plain juvenile. Yet some of these stories, in print and on the air, were actually written by people who have never been online at all, or whizzed through for the predictable "my first two weeks online" tour we're all so weary of reading so they could submit juicy, clever-sounding copy by deadline.

*The reality is that the online world is like our real world in more ways than not.*

Indeed, there are times online when you feel like a visitor from another planet who is seeing the array of human behavior in all its variations and aberrations for the first time. It's like reading Samuel Butler's book, *Erehwon* (*nowhere* spelled backwards), about a land in which mental illness gets sympathy but physical illness is considered a weakness, and everything else is the opposite of the way it works in the world we know. Nonetheless, the reality is that the online world is like our real world in more ways than not.

Regardless, you owe it to yourself to find out what's it's really like and make up your own mind. As Lisa Kimball of The Meta Network says, "There's a lot of hype, but anything with that much hype is something you need to know about."

# The nature of the culture

In one sense, there's no such thing as online culture, because what's acceptable on one network isn't on another, and what's the norm on a commercial service isn't on the Net. The more the commercial services and the Internet converge the more the norms will merge, but they will still always be somewhat distinct, as will what's acceptable and what's not within a given system. So one of the first things to learn is "when in Rome. . . ."

The Internet was created to keep communication open amid anarchy, in the event of all-out nuclear war. So that is its fundamental nature: The Net thrives on what some would consider anarchy. The commercial services, on the other hand, were created by businesspeople to make money, and that will always be their fundamental nature. The problem arises when you try to convert the Net to a place to make money, and open gateways between the two so people used to one culture mingle with people used to the other. That's what is happening now, and the inevitable result is culture clash. It also makes the whole online world one of the most interesting social experiments humankind has ever experienced, although nobody intended or expected that.

# What's highly valued by people on the Internet

Even though the Internet was created by the military branch of the government, it was shaped by scientists, academicians, and students, most of whom hold certain values very dear. In fact, it wouldn't be too much of an exaggeration to say that the Net was shaped and still is run, to the extent that it's run by anyone, by idealistic youth, aging hippies, and people protected from the harsh realities of the real world by tenure and group insurance and pension plans. Many have free accounts on the Internet or have never paid more than 20 bucks a month to roam at will. All of that influences their perspectives and comes into conflict with those who have other interests to protect. That's not to say that their values aren't admirable because, in general, they are. The following are the basic ideals of the die-hard denizens of the Net:

*Freedom of expression* Read *unfettered.* Say whatever you want, as long as you can get by with it without being run out of the group by other people. Censorship in any form is abhorrent to most people online, even though they realize that the trade-offs are sometimes undesirable, just as those same trade-offs are in the real world.

*Privacy* The stance here is that "Nobody but nobody reads my e-mail!" Not the government, even if they claim they're only on the lookout for terrorists, not my boss, not my spouse, nobody. Period.

*Equal access* Translation: "The Net should be free or at least affordable for everyone." This sounds good, but even the biggest zealot of free access would concede that someone has to pay for the computers, the fiber-optic cable, the staff, and all the rest that it takes to maintain the network. That someone, in the U.S., has been the taxpayers, but the government won't be footing the bill anymore so the Net must change to accommodate commercial sponsors.

*Free information* "Information wants to be free" is the common saying on the Net. Maybe so, but because of the current transition to more commercialism, the trend is to pay per view or pay per download. The outcome of that is anyone's guess.

*No commercial mention* Tout your product or service blatantly in a discussion group, either on the commercial services or the Net, and you'll get flamed or censored. Yet it's okay for advertisements to pop up at the bottom of your screen every time you hit the Return or Enter key on Prodigy, or for Nabisco to create an entertaining soap opera to run in the America Online version of the *New York Times*, even though it's really a promotion and not just entertainment. And it's okay for anyone who has the price of admission to put up a Web site promoting whatever they want, as long as they can find a service provider that will allow it on their computer and nobody gets busted.

*Contribute, don't just take* If you only lurk (read messages without participating in the discussion or download information or software programs without ever uploading something to contribute to the body of knowledge), some consider that being rather like a leech. Others defend the right of people to lurk as their prerogative. I admit a strong antilurker prejudice, because if we all merely took and gave nothing back, soon there would be nothing for any of us to take. So the tacit understanding is that you freely share what you know, at least as much as you partake of what others offer, and that if someone helps you, you either reciprocate or pass the favor on later when someone else asks you for help. Would that the world always worked that way.

*Reason and logic* This might sound like a contradiction to Net veterans who've witnessed many flame wars, as intense arguments characterized by personal attacks are known, and seen how irrational and emotional people can get amid them—again, just as they do in real life. But this is mainly a mind-to-mind medium, and the majority of people online are well educated, so being able to present a logical argument and support it with verifiable facts or defensible reasoning still makes points.

# The paradox of elitist attitudes

It's rather ironic that some of the very same proponents of the high-minded values stated previously exhibit some of the strongest prejudices against others, such as these:

## The right address

Most Net idealists would bristle at any innuendo that if you don't live in Brentwood or Beverly Hills you somehow have less to offer than those who do. Yet they think nothing of denigrating those whose online addresses end in aol.com, or anything connoting that you use a commercial service for access. The America Online addresses take the most abuse, because AOL grew so fast and unleashed a horde of novices on the Net who, when they knew no better, violated many rules of Netiquette. It will be a few more years before they live that down.

There's even a none-too-subtle prejudice against those who use networks with proprietary software that automates much of the process, such as Netcom's Netcruiser or PSI's Pipeline. Therefore, if your address ends in (as two of mine do) @pipeline.com or @ix.netcom.com—the "ix" part of which is a tipoff that you're using Netcruiser—you're still setting yourself up for this silly snobbery. The ultimate, of course, is to have an address that includes your own name, which is like having a vanity license plate.

## Ostracism of newcomers

Among members of Alcoholics Anonymous or Narcotics Anonymous, there's a kind of checking-each-other-out process that occurs, much as dogs sniff one another when they meet, to see who has been clean and sober longer. In Utah, one of the first questions people ask you even at social gatherings is whether or not you're "LDS" (a member of the Church of Jesus Christ of Latter Day Saints, or Mormon). This is very similar to what often goes on online, because it's not uncommon for someone to say, in a tone so disdainful you can almost hear it, "How long have you been on the Net (or 'online')?" Implication: "You're obviously a 'newbie' and don't know from up, but I do." If you have a .com address (indicating that you use a commercial service gateway rather than a direct-access provider), they usually assume you're a newbie even though many people use more than one service now. I often log onto the Net from America Online, CompuServe, or Delphi, even though I have Netcom and Pipeline accounts, and the negative reactions and snobbery of some people are unmistakable.

# The nature of people online

As for what happens between people online, half of what you've heard is hype and the rest is human nature, so let's put it in perspective.

## Who's online

The people who use the commercial services are predominantly middle-class, well-educated, white men, which isn't representative of the population in the real world. Based on reports from the online services and industry analysts, daily online experience on several services and the Internet, and allowing a fudge factor for women who use online accounts in their husbands' names, my best guess is that as many as one-third of the people online are women, as shown in Fig. 1-1. (Thanks to Jupiter Communications, Link Resources, and SIMBA for their help in verifying these percentages.) Excluding Women's Wire because of the way it skews the results, the average is 26%. Including Women's Wire, the average for all services cited is 33%. John Quarterman, a respected research demographer with Matrix Information, estimates that a 2:1 ratio holds true for the Internet too.

*They're warm, witty, spontaneous people*
*who are usually very bright.*

Because you can't see the other people, other than through photos if they choose to upload them, you're not aware that you're outnumbered if you're a woman. But if you look at it another way, when you're online you're among sharp, savvy people who are adventurous enough to try something new before the average person.

"I find it a place of incredible, wonderfully supportive people and a wealth of knowledge," says Delynden (Dee) Lersch, of Dallas. "Like any crowd, there are a couple of bores and buffoons. But, as a social set, they beat the people you'll

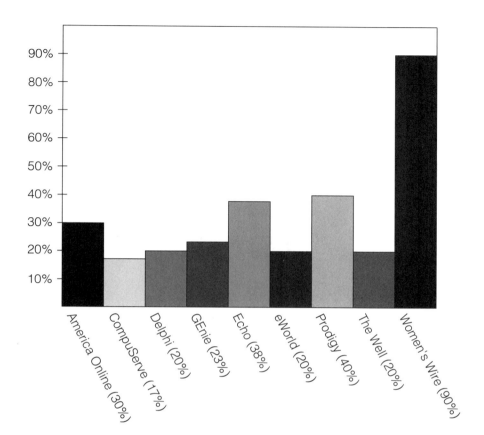

**1-1** *The percentage of women on the commercial services as of early 1995.*

find in any bar and in most PTA meetings. They're warm, witty, spontaneous people who are usually very bright."

Or as Connie Strand, of Kansas City, Missouri, sees it: "It's fun and it doesn't cost that much to give it a try. It's like a window into people's lives. . . . Around Mother's Day, people were writing [public] letters to their mothers who were deceased, and there were letters from adopted children 'to my mother, wherever you are' or . . . thanking their birth mothers. The feelings that came out were quite moving. These are people who are articulate, and it was really interesting to hear how people really think and what's important to them. Sometimes it's like being a mouse in the corner of a therapist's room. . . .

You can send this message out into cyberspace, and you're touching whoever happens to read it."

# Bozos and jerks

The proportion of bozos online is roughly equivalent to the proportion of bozos anywhere. It's just easier to get rid of them online. Many online services even have what's commonly known as a "bozo filter" or "squelch feature," which is a way you can block all messages from anyone you choose with the click of a couple of keys. And there's always the Delete button. You can also block all instant chat invitations.

## How to block instant chat invitations

*America Online* Click on Members on the main-screen menu bar, choose Parental Control, and choose Block Instant Messages. Can be done only with your Master Account or the screen name you used when you joined the service.

*CompuServe* If you're using the CompuServe Information Manager software, you can click Ignore whenever anyone sends you a chat invitation, but you can block only one message at a time, not all of them at all times.

*Delphi* Just type /busy at any command prompt after you log on. This blocks instant chat messages for that session only, and might change when they introduce new software in 1995.

*Echo* In your profile, which you reach by typing .profile, type yo no, short for "no instant messages" (Echoids call them *yos*). You can reverse it with yo yes.

*eWorld* Send a one-on-one message addressed to $IM_OFF, with anything in the body of the message. This blocks all one-on-one invitations. To reactivate, do the same, substituting ON for OFF.

**GEnie** When someone sends you a chat invitation, you'll see a number beside that person's name. In manual mode, type /squelch followed by the number. GEnie is introducing new graphical software in 1995, so that command will be different with the Windows or Mac software.

**Interchange** Ask for instructions in Member Support.

**Prodigy** JUMP Chat —> Setup. Highlight the name of the member of your household whom you want to be unavailable for instant chat messages, click on Clear, then click OK. The individual has chat area and instant-chat access only if it says "currently enrolled" after the name. Once you do this, it should be blank, although you'll still see the member's name.

**The WELL** In your profile, which you reach by typing .profile, type mesg n, which means no instant messages (WELLbeings call them *sends*). Reverse it with mesg y. That's for manual mode, and won't apply to their new graphical software.

**Women's Wire** In the Edit menu, set Preferences to Don't Accept Chat Invitations. You can also leave your options open, but click Decline when someone invites you to chat, or just ignore the invitation.

# Gender benders

Gender benders are men and women who get their kicks from masquerading as the opposite sex, particularly in chat areas or forums where the nature of the discussion is about sex, relationships, or personal problems. It doesn't necessarily have anything to do with their sexual preference. Sometimes they're just curious about reactions, sometimes they get a thrill out of the deception or role-playing. The truth usually comes out, leaving people who bought into the ruse feeling embarrassed at best and betrayed and emotionally wounded at worst. (For more on this, see chapter 6, *Sex and romance*.)

# Online discrimination

Time and time again women say that two of the things they like most about communicating online are that men can't interrupt them when they're speaking, as so often happens in face-to-face or even phone conversations, and that because no one can see them they can't be judged by their appearance or gender. People can upload photos of themselves to forum and chat-area libraries, but that's optional. Normally, whether you're male or female, gay or straight, beautiful or ugly, fat or thin, young or old, able-bodied or not, everyone's the same online.

# Real names vs. screen names

Fake names, online or off, encourage people to take advantage of that anonymity and say things they wouldn't have the nerve to say if people knew who they were, or to otherwise act in reprehensible ways. "Handles" have been an online tradition since the days when public-access online services meant some hacker running a BBS out of a corner of his garage. But that no longer works. I could go into a whole social analysis about why, but what's relevant here is that people are more likely to make hostile, lewd, and just juvenile remarks when they can hide behind a mask. This, combined with the sensationalized media reports of online sexual harassment, has made many women afraid to use their real names. So some sign on with their husbands' names or make up a handle that's ambiguous.

There's no evidence to suggest that using a gender-neutral name makes any difference one way or the other in preventing some jerk in a chat room from zapping you a message saying, "Hey, babe. . . ."Although if you call yourself SexySue or SassyLassie online and hang out in Flirts' Nook or Romance Connection, and you get propositioned, you forfeit your right to complain about it. So one important consideration in creating an atmosphere where people treat people as people, rather than type on a screen, is to have and enforce a policy of real names only, preferably one person, one name. This is particularly important for most women to feel comfortable.

There are situations online, such as chat rooms or forums where people discuss personal matters, that pseudonyms are appropriate, however. As Amy Bruckman, a grad student and researcher at MIT says, "Dress for the occasion."

## Sexual harassment and stalking online

Frankly, the flap about online sexual harassment is greatly exaggerated. Yes, people do occasionally send suggestive or even lewd messages, but they're almost always confined to or the result of interactions that began in chat rooms, which are the online equivalent of singles bars. Completely ignoring them works best, but a reply along the lines of "not interested" or "get lost" works too.

Actual stalking, as defined by the law, is all but nonexistent. Some newcomers, paranoid from all the press hubris, get upset if someone they've talked with only in a public forum sends them a private e-mail message, or confuse stalking with what most online veterans would call flaming. Says Fresno Police Detective Frank Clark, who investigated one of the most notorious alleged online stalking cases, "Most of what's going on is offensive, but not an offense."

As Paula Span wrote in a 1994 essay on women and computers for the *Washington Post Magazine*, "Somewhere between enforcing 'nice networks' for women and having women set upon by wolfpacks roaming the Internet, there must be a workable middle ground. I count myself among the optimists, partly because there are systems that demonstrate the possibility of egalitarianism. It doesn't happen by accident, but it does happen."

# The nature of online communication

It has become fashionable to talk about differences in ways men and women communicate, online or off. One study showed that men tended to steadily escalate the ways in which they tried to inhibit participation by women in a discussion if more than 30 percent of the comments came from women. However, that study was done during a discussion of sexism, which certainly gets different reactions from men than a discussion of gardening would, thus the often-quoted results are skewed and misleading. Another study showed that men tend to dominate discussions online even when the male-to-female ratio of participants in a given group is roughly equal. To the extent that this is true, it reflects the way women have learned to behave in life in general as much as it reflects anything about the online culture.

Women lurk far more than men, meaning they don't speak up as much online. Women are just as free to post and reply to messages as men are, so if they don't it's their choice rather than something imposed upon them. In interviews, women have told me that they often don't make comments because they're afraid of looking silly or stupid, which is much more a concern of women than of men. But, again, that's a cultural phenomenon born and bred in the offline world, simply carried over online, and a matter of personality rather than persecution.

## Male vs. female communication styles

Susan Herring, a linguist at the University of Texas at Arlington, says, "I think that women are intimidated from participating because there's a lot of flaming on the Net. It doesn't have to be directed at the women. Women are intimated by that more than men are, although men are too. Men would say, 'But that's the way it is, and I think it's fun as long as they're not shooting arrows at me,' but women have strong feelings of aversion."

*Women are not into conversation as a contact sport.*

Herring says there are telltale signs that can reveal when the writer is a woman, even if she's masquerading as a man online or using a gender-neutral name. "Little girls are encouraged to be nice and take other people's feelings into account. . . . Extremely polite behavior or expressing appreciation is female, and extreme behavior is overwhelmingly male. Women hardly ever flame. . . . The women send pure, unevaluative information, whereas men express opinions and argue with each other. The women's messages take the other person into account more."

As Lisa Kimball of The Meta Network puts it, "Women are not into conversation as a contact sport. Some men seem to find it fun and recreational to have political debates to see how cleverly they can state a position, whereas women are more likely to approach the debate from 'How can I understand what you're saying better?' They don't enjoy somebody coming up with a good zinger as much. It just isn't interesting or fun."

## Computer communication can make people more candid

Tim Finholt, an organizational psychologist at the University of Michigan, says "Computer groups have been criticized for being impersonal and composed of unknown or barely known members who can't relate to each other as human beings. But in fact, computer mail flattens the barriers of time, distance, organizational boundaries, and, in some instances, hierarchy. The absences of a time and setting for the group meeting and the virtual invisibility of individual members seems to free people up. . . . You have the illusion of belonging to a small group, yet the recipient of the message is not responding to it while surrounded by others in the group, so the power of peer influence or groupthink is diminished. Consequently the recipient is more likely to reply honestly and from the heart."

## What should save time takes time

"A lot more inconsequential stuff is left to e-mail, and people ramble on and on," says Mark Schulman, Academic Dean at Pacific Oaks College in Pasadena. "This is certainly the case with the Internet, since it's an ultra-democratic experience; you just don't have much sorting going on."

# Things the manuals rarely tell you
## Password protection

Each online service has different requirements about how many letters or numbers a password can contain, whether it can contain any numbers or nonletter characters at all, or whether it must include *only* numbers and nonletter characters. Experts advise avoiding any words that are in the dictionary and any numbers that people might guess, such as any part of your Social Security number, birthdate, or graduation year. Also, use different passwords for each service even if they allow the same one, just as you use a different key for the locks on the doorknob and the deadbolt.

Bank machine codes were bad enough, but as soon as you're online you'll have so many more ID numbers, screen names, passwords, and subscription e-mail addresses that remembering them all is nearly hopeless. Worse, when you subscribe to an Internet mailing list or publication on the Web, you'll automatically get a long return message telling you how you subscribed and how to unsubscribe. Yes, *unsubscribe* is usually the term, grammatical or not. Most people just save those in a file, and that can work if you're conscientious about always filing things alphabetically. If you aren't, like most of us, digging through the file to find the alt.whatzit newsgroup instructions is as frustrating as being bombarded by information you don't want anymore.

 So write the names, ID numbers, and passwords down, but keep them in a safe place. For the newsgroups and mailing lists, get an ordinary address book, use it just for Internet addresses, keep it within arm's reach of the computer, and enter the name of the list or group, how to get on and how to get off. Life will be easier. The address book is also handy when someone asks you how they can join something you recommend.

## How to stop what you started

If you started some procedure online and want to stop it, but it won't stop, on PCs you can hit Ctrl–C, which cancels the action, or Ctrl–Break (Break is also the Pause key on the upper right corner of PC keyboards). This will usually put you back at the beginning. On America Online, unlike other services, if you start to download a file and find out it's taking too much time or isn't what you wanted, just hit the Escape key and that's that.

## What to do when you want to quit

On services that are all text, without graphic icons that help you navigate, you can usually find your way out and offline by trying Exit, Off, Bye, or Quit.

## Copyright violations

Contrary to what some believe, it's not all right to repost an article you find online or publish it in your company or organization newsletter, even with full credit to the original author and publisher. You can quote only small portions for some purposes, such as commentary, but the legal definition of small and for what purpose is ambiguous and depends on the nature of the article.

Otherwise, if you republish it, including online in any way, it's a violation of international copyright law subject to sizable fines and legal fees.

# Netiquette
## Don't waste bandwidth

Don't leave one- or two-word messages such as "Thanks" or "I agree." If someone has already said what you might have said, let it go. If you want to thank someone for a favor, do it in e-mail, unless you have something else to add.

## Watch your language

DON'T SHOUT! Typing in all caps has the same effect as SHOUTING. And remember that the only impression people have of you at first is how you type. Until recently, when more net software programs incorporated spell-checkers, people overlooked obvious typos or misused words. No longer. Clean, coherent copy is now considered the equivalent of being neatly dressed. Some will judge you by how well you write, but you needn't be a Pulitzer candidate. Just watch your grammar, spelling, and punctuation.

## Observe and learn the norms

Lurk a while until you get a sense of the tone and content of the discussion. However, the norm on all networks is that anyone can join in any discussion, so you needn't wait to be invited. They can't see you, so if they don't know you're there they can't invite you. This isn't a time to be shy. If you are truly shy, here's your chance to venture forward in a safe, low-key way and see how it feels.

## Practice before you preach

There are free new-user and help forums on the commercial nets and counterpart newsgroups on the Internet where you can practice posting and downloading messages and files until you're sure you're doing it correctly. You'll save yourself embarrassment and avoid irritating others if you use them. Once or twice through the routine, and you'll have it down pat.

And remember, the forum or newsgroup didn't come into existence the day you arrived. Wait a few weeks until you fully understand what's going on before you start making suggestions about how things could run better.

## Just the FAQs, m'am

 Read the FAQ, or Frequently Asked Questions files available for most newsgroups and some libraries on commercial services so you don't ask what a thousand people have already asked before.

## Real people, real emotions, real reputations

Throughout this book I say "in the real world," but I mean it partly tongue-in-cheek because the online world and, more importantly, the people there are real. When you're sitting and typing silently, alone, seeing only the words on your monitor, with none of the immediacy and reactions or nonverbal feedback you get face-to-face or even by phone, it's easy to forget how words can mess with people's minds. No one past the first grade believes "Sticks and stones can break my bones, but words can never hurt me." The people online are real and you're in public, so please never, ever forget it.

Because 70 to 90 percent of communication between people is nonverbal—the glint in the eye, the shifting in the chair, the slight smile, the derisive sniff—everything depends on your words. So choose them carefully and consciously, and remember that irony and sarcasm, especially, don't come across well in pure type, which is why it's easy to hurt someone's feelings unintentionally. That's also why people use emoticons, those corny, cutesy symbols that substitute for the nonverbal cues, such as smileys :-) or grins <g>. Even "just kidding" is fine. Better to sacrifice some subtlety than be misunderstood.

## Common courtesies

If you mention a good newsgroup or forum or file that's online, tell people how to find it themselves, in the same message. Try to address people by name, as in, "Jane, you mentioned that. . . ." It adds a friendly feeling, and helps you get to know people. Introduce yourself when you're new to a discussion group and intend to participate frequently. Be helpful and follow the Golden Rule, and you'll do fine. Really.

# CHAPTER 2

# A guided tour

**Y**ou don't need to be a computer whiz to whiz around online. When I first logged onto CompuServe, I knew how to use the basic functions of a word processing program, nothing more, and had used computers for only about four years. Despite that, a year later I became a sysop (short for systems operator and pronounced *SIS-op*). Any of my techie friends would tell you that I still don't know much about how all of this stuff works, nor is learning more than I need to know high on my priority list. All any of us need to know is enough to make it work *for* us. Anytime we want to know more, it's easy to find out online. I have never, not once, had any question about any subject that I couldn't find an authoritative answer to online, usually within 48 hours or less. What's more, I now have friends and co-workers across the country and in other countries, and opportunities and resources at my fingertips at any hour, that were only fantasies before. And so can you.

The only assumption here is that you know the basics of how to use a computer and that you have one, are about to get one, or have access to one. If you're still skeptical and are just shopping, this book will help you decide if online services are for you, and which ones you'd like to try first. If you're already online but are still a novice, or have been online for a while but want to know more, this book will help you too. You might even be surprised at how much you can do online that's useful in real life.

# A typical online service

Let's start by getting an idea of what online services look like, in the abstract, and how they function, which will help if what you know about online networks so far is still just confusing jargon. It will also help you understand more of the benefits and drawbacks described in the next chapter. It's easiest to talk about the commercial services such as America Online, CompuServe, and Prodigy first, because they're all pretty similar. So what follows applies mostly to the commercial services, rather than the Internet. Then in the next section we'll look at how the Internet is similar and different.

Okay, please clear your mind of all preconceptions. Imagine that you're attending a convention held in a big building. The first thing you do when you arrive is register, right? That's the same thing you do when you log on (connect by modem) to an online service for the first time. The next thing that happens at a convention is that someone hands you a packet of information that tells you what's going on and where. Maybe you read the directory in the lobby too. You'll find essentially the same thing online, except that the directory is on your computer screen.

On the directory or in the conference information packet, meetings or seminars are usually grouped by categories; online, information is also arranged by category. So the first screen you'll see shows broad categories, such as Business and Professional Interests or Home and Family. You can explore all of them eventually, which is the best way to find what interests you most or to discover things that aren't obvious at first. But you have to choose a general area to begin with, of course.

Choose any of the categories, and it will lead you to more specific choices. Some of those are places that offer information to read or use in some other way, but some will take you to places where people carry on public discussions by posting messages and replying, back and forth. There might be hours or days between one person's comments, but others respond or bring up new points meanwhile. Sometimes, to avoid the lag time, two or more people carry on discussions in what's known as *real time*, meaning right now, but still by typing messages to one another. Behind the scenes, people are also exchanging private messages, just as people pair off for private conversations at conventions.

– 21 –

# The five basic features of an
# online service

The choices mentioned in the previous section encompass five main features of a typical online service:

➤ Forums (bulletin boards)

➤ Libraries

➤ Real-time conferences or "chat mode"

➤ E-mail

➤ Databases

# Forums

Just as at most conventions, there are simultaneous sessions with presentations and discussions. The next step is to choose which one you want to attend. Or you can do what many of us do at conventions, which is to sit in on one session for a while, then wander over to another so you won't miss what people are talking about there. Like convention programs, these sessions are set up by topic. Online, they're known as *forums, conferences, roundtables, message boards*, or *bulletin boards*. By whatever name, they're all the same kind of thing. Technically, they're electronic bulletin boards, although on the big commercial networks such as CompuServe or America Online they're usually called *forums*. The term *bulletin board system*, usually written BBS, is more often used to refer to local or specialized services that are much smaller than the major online networks, yet similar in how they operate. I'll introduce you to a couple of good ones in the next chapter.

Each forum focuses on a particular topic, for instance, health and fitness. Within each forum, there are several different subtopics or sections that usually remain the same. In the Health and Fitness Forum, you might find sections on exercise videos, diets and nutrition, and a dozen or more other subtopics. Within each of those subtopics, members start whatever discussions they want to, thus topics change frequently. A hot topic can generate discussions for weeks, even months. Others might flare and fizzle because they're related to some transient event or just don't appeal to many people. Although you see only one tier at a time on your screen, the overall structure is much like a classic outline (with I.A.1.a levels), roughly like this:

```
                    Health & Fitness Forum

Exercise Videos    Diets & Nutrition  Eastern Medicine   Ask Dr. Whozit

Discussion Title   Discussion Title   Discussion Title   Discussion Title
Discussion Title   Discussion Title   Discussion Title   Discussion Title
Discussion Title   Discussion Title   Discussion Title   Discussion Title
Discussion Title   Discussion Title   Discussion Title   Discussion Title
```

Another way to think of it is in layers. Each choice you make from a menu leads you to the next layer down, or deeper, to more detail. This is rather transparent on some services, because you'll see right-pointing arrows to go forward, or deeper, and left-pointing arrows to go back to the previous level. However, your screen usually shows you only where you are in the hierarchy at the moment. Don't panic, because you can always retrace your steps if you feel lost in the woods, and the computer will remember where the bread crumbs are. So the actual screen you see will look more like the one shown in Fig. 2-1 (although I've superimposed two here).

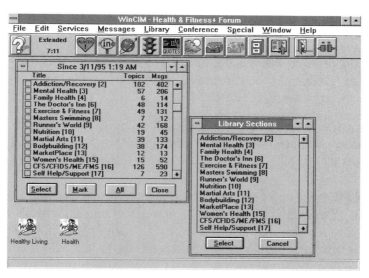

**2-1** *The Health and Fitness Forum on CompuServe.*

Got it so far? Good. Now let's go back one level, to where we were: online discussions. There are only two aspects of communicating online that are different from any group of people getting together and having similar discussions offline somewhere like a conference or a café. First, there's a written record available of what everybody said, so that someone who comes in later can easily catch up. Second, anyone can join the public discussions and chime in at different times on different days, from many different cities or even different countries. So the discussion isn't over just because the workday ends at 5:00, or because the clock strikes midnight and it seems impolite to stay at your friends' house any longer, or because some people involved in the conversation leave. You can start a new topic yourself whenever you want, although it's best to just read what other people are saying at first, until you get a sense of the tone and level of the discussion and what has already been said, just as you would if you dropped in on a gathering anywhere offline.

# Libraries

Most but not all forums also have libraries. They generally mirror the section topics and contain three basic items:

➤ Documents, all in generic ASCII text so people with different kinds of computers can use them.

➤ Graphics or photos converted to computer files that require special tools to see what's inside (the tools are available online, or even built into the software that some services use).

➤ Programs, each of which specifies what kind of computer it will run on, generally a Macintosh or PC (IBM-compatible). If it's for a PC, it will say whether it's for DOS or Windows.

Everything in the libraries is available for you to download into your own computer, and it's all free. Software might be only demos though, so you'll need to buy the full programs—or, in the case of shareware (try-before-you-buy software), pay to register them if you keep using them.

*Download* means to copy a file or message down into your computer. *Upload* means to copy a computer file or send a message up into cyberspace, where it ends up stored on the online service's computer for other people to download. Think of cyberspace (coined by William Gibson in the novel *Neuromancer*) as some great place out there in the sky, which it is, kind of. Upload *to* cyberspace, download *from* cyberspace. How to upload or download is another matter, but there are instructions on the screen, appropriately called *prompts*, that guide you through what to do next. You can practice in free areas online or do something radical such as read the software manual. However, most software has a help screen that explains it because software people know most of us never read the manuals.

# Real-time conferences or chat mode

The key thing these have in common—whether happening within a forum or in separate areas—is that they all occur in what's known in computer jargon as *real time.* This means that two or more people type messages back and forth to each other while they're actually connected online, as opposed to leaving a message on a forum's bulletin board, then logging off and checking back later for responses. If you're old enough to remember the term, it's like a party line, except you communicate through typed messages rather than spoken words. Sometimes chats or conferences are very informal, with just banter or get-acquainted talk. Other times the sysops invite a special guest to be a presenter and it's more like a formal seminar. However, hardly anything involving inter-action between people online could be called formal, given human nature and the *laissez-faire* nature of the medium.

# E-mail

This is one of everybody's favorite and most-used online features. E-mail is available on all online services, and writing, sending, receiving, and reading e-mail are some of the easiest things to do. Given the right service or software, you can also write and read messages offline without being connected, thus without paying for online time. You can exchange e-mail with anyone on any online service anywhere, and *anywhere* includes about 200 countries world-wide. Fair warning: it's so convenient, saves so much time, and can be so enjoyable that it's instantly addicting.

"It's hard to explain to someone who's never used e-mail why it's a good thing," says Lee Sproull, a sociologist, professor, and researcher at Boston University's School of Management. "People say, 'I can make phone calls or send postal letters, so why would I need this other thing when it's just like things I already have?' Unless you've done it, you don't understand why it isn't like anything else. It isn't simply a substitute, it's its own thing."

Like everything else online, e-mail will seem more familiar than foreign to you, yet there are twists that make it, as Dr. Sproull says, its own thing. In some cases, it costs you nothing in addition to your basic monthly fee. In others, the charges are so nominal that it costs less to send e-mail than regular mail through the postal service. Better yet, what you send arrives almost instantly, and you often get an answer the same day, which is why ordinary mail is also known online as "snail mail." E-mail is a mainstay for businesses and, increasingly, for families who use it to keep in touch with far-flung relatives (see chapter 4, *Friends and family*). You'll soon consider the several steps you have to go through to send mail the old way or playing phone tag to be real hassles and time-wasters. But don't make the mistake that many do of using nothing else online but e-mail because, as terrific as it is, it's only one part of the rich resources online.

## Databases

There are different kinds of online databases but, again, they're all essentially alike. Chapter 18, *Online research basics*, gives an overview of how to use them. They're all collections of information, typically in a generic form of straight text without any graphics (known as ASCII), which makes them usable on any computer. The great thing about the databases, aside from the wealth of information they contain, is that you can search them in several ways just by tapping commands on your keyboard. Those commands are often as simple as ordering from a menu, because that's exactly how they're organized: by menus. Databases and forum libraries, described earlier in this chapter, are the equivalent of the information tables at conventions where you can pick up or buy whatever you want and take it home with you.

So now you know what a forum is and how it works, what libraries and online databases are and what's in them for you, and how you can communicate with other people in forums, chat mode, and conferences or by e-mail. You also know what *upload*, *download*, and *real-time* mean.

In addition to all these things, there's typically an array of other services ranging from stock quotes and newswire reports updated every half hour to horoscopes. Some are covered by your monthly access fee and some aren't, but it's easy to tell which are which. Just be careful that you don't sign up for a free trial period that automatically converts to a permanent account for some extra service, and starts showing up on your credit card bill. Chapter 12, *100 ways to save time and money online*, covers some of the best time-saving features available on various services.

## Similarities and differences on the Internet

Despite hundreds of news and feature stories about it in the last year alone, nearly 58 percent of the white-collar professionals surveyed in a late-1994

Gallup poll for MCI said they hadn't heard of the Internet. Oddly, two-thirds said they *had* heard about the information superhighway. Rather than contradicting each other, MCI says that these results more likely indicate that people simply don't understand the connection between the two terms. However you interpret it, it's astounding. If you bought this book, you're not likely among those who are still completely in the dark, but you might be among the 83 percent of women who still don't know how to access the Net.

To set the record straight, *online* is not synonymous with *Internet*, because the term *online* also includes commercial services such as Prodigy or a local-only BBS or freenet. The *information superhighway* or *data highway* or *national information infrastructure* are all essentially different names for the same idea, but none of those are synonymous with the Internet either. Although the Internet is as close as we've come to that more far-reaching vision, it's only the foundation. Things are changing rapidly, and will probably change enough in the next decade to make what we have now seem rudimentary.

In case you don't know but are remotely curious, here's the gist of what the Internet is: The U.S. government launched the forerunner to the Internet during the long, drawn-out Cold War as a way to bypass blackout zones in case some of them were nuked, so they could still communicate with the elected officials and the troops. It evolved into a way for scientists and academicians to communicate and collaborate, and now it's evolving into a way for you and me and all of us to do the same. That's the shortest history of the Net I've ever read, much less written, so if you want to know more, read almost any of the online guides listed in the Bibliography.

> *Cyberspace is not weird, mysterious, or scary. . . .*
> *To get there, all you need is the right equipment*
> *and the right attitude.*

At this writing, there are nearly 200 countries on the Net. But here's a key point: They aren't really "on" anything, because the Net is more of a concept, just as cyberspace is a concept, or a "hypothetical construct" as scientists would say. Neither exists in reality, although once you find places online that you like to visit frequently, you begin to think of them as real.

The people you get to know there are real, regardless. If you're on the Net—or "wired," in the argot—that merely means you're able to connect to everybody else who is able to connect, which means everyone who has a phone line, a modem, and the necessary software to tell the modem what to do. Including you. Cyberspace is not weird, mysterious, or scary. You already go into cyberspace every time you talk on the phone, so it's no big deal. It just sounds cool because it's still so new to most people, even though the Net has existed for almost a quarter of a century. To get there, all you need is the right equipment and the right attitude. You've already got the phone line, maybe even the modem, and you wouldn't be reading this book if you didn't have the right attitude. So you're already halfway there.

In sum, the Internet isn't the name of any service because it's a network of networks, each of which connects or disconnects as it chooses. (*Net* usually means the Internet, and net without the capital N refers to any other service connected to it, such as CompuServe or America Online. Technically, it's all the Net, but it gets too confusing to be such a purist.) There are millions of people using computers who can send and receive information via the Net. If it were possible to photograph the Net over just the United States, it would look somewhat like what's shown in Fig. 2-2.

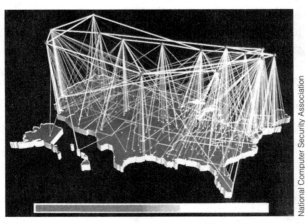

**2-2** *Computer-simulated photo of U.S. Internet connections in 1993, before the boom of the past two years.*

Because no one set all this up with a common model or standard, as they did with the commercial services (America Online, CompuServe, Prodigy, and so on), the Net is much more free-form. Chaotic, some would say. It constantly changes, but there are general attributes that are consistent. Most of those bear some resemblance to the commercial nets, with variations. Let's look at them in the same order as the areas of commercial services earlier in the chapter, although the names vary:

## Newsgroups (bulletin boards)

The closest thing to the forums of commercial services on the Net are news-groups. As on the commercial services, each one is about a particular topic, and they're composed of messages that are posted so others can read and reply to them. As with the forums, anybody can start a discussion within a newsgroup, and anybody can contribute to that discussion. On the Internet, however, you can also start a whole new newsgroup, as long as it survives the process of calling for consensus on whether it's a go or no-go. Most of them do.

Without some system of organizing this, it truly would be chaos, so there are only a dozen or so basic kinds of newsgroups, each of which is identified by abbreviations that conform to syntax that computers can handle. The most popular ones are soc (society or social issues), misc (miscellaneous), alt (alter-native or more off-the-wall topics), rec (recreation, hobbies, the arts), news (news-related issues), comp (computer stuff), sci (science) and talk (debates and diatribes). To be more definitive, you simply make up whatever else gives people a clue as to what the newsgroup is about. Thus you get names such as misc.kids and rec.music. Easy, right? But that's too broad, so they add on more things to explain. That's how rec.music splits off into rec.music.classical and rec.music.jazz and rec.music.bluenote. Because anyone can start any kind of group, within legal limits and not always within the bounds of good taste or common sense, you end up with groups like alt.tasteless and alt.barney.die.die.die (yes, the purple dinosaur).

You get the idea. There are somewhere between 7,000 and 9,000 news-groups—new ones start and others fold frequently, so it's hard to keep track—on just about every conceivable topic. Many are redundant or overlapping or too technical to generate wide interest, but that still leaves thousands of good possibilities. Of course if you can't find the one you want, you can start a new one, but wait until you've thoroughly checked through what's already out there. With daily diligence, that could take months.

## Archives

Newsgroups don't have libraries in the same way forums do, but they do usually have archives of discussions, directories of members, and such. Learning how to use them isn't for novices, though. You have to know where they're stored and you also need to tell the computer where they're stored, how to get there, and what you want it to find and send to you.

## Internet relay chats

Also known as IRC or sometimes just "talk," this is the real-time conference or chat mode on the Net. It's the same basic thing as on the mainstream nets, but based more on the CB radio model. Therefore, there are different channels rather than "rooms." Calling them channels makes a lot more sense, given that this is communication through the airwaves. But a rose is a rose.

## E-mail

Ah, something that's actually the same thing, by the same name! Don't you feel right at home already? It even works the same, although the addresses are longer. With the addresses, too, some consistency exists to create order out of chaos. There are three keys:

➤ The address of the person, called the *username*

➤ The address of the specific computer where that person's account is stored, called the *domain name*

➤ The address of that computer, known as the *zone.*

There are only a few categories of zones: com (commercial service), edu (educational, usually a university), org (organization, usually a nonprofit), gov (government agency), int (international agencies), mil (military), net (network administrators). In between the username and the zone are other parts of the domain address, sometimes. The following, for instance, are two of my Internet addresses:

judith@ix.netcom.com
broadhurst@aol.com

When you're new, it all sounds totally cryptic. In fact, it sounds totally cryptic to most of us no matter how long we've been around, so don't let it put you off. Just remember that there are never any spaces, there are a lot of periods, and try not to make any typos.

## Mailing lists

One of the best things on the Net is a variation of e-mail called mailing lists or listservs. Don't worry about the jargon; the thing to remember is that listservs, mailing lists, and lists are all the same thing: group e-mail. However, they're also a lot like newsgroups because they're on a particular topic. The way it works is that people send messages to the computer that stores the mailing list, which is what sends the mail out to everyone, thus it's the *listserver*. Then people reply by sending mail back to the listserver, which sorts it and sends out the replies so everyone can keep up with the correspondence.

Some lists are moderated, meaning a real person sorts through the mail, takes out the extraneous or redundant information, and sends out edited versions or compilations. The purpose isn't censorship (which is considered a heinous crime on the Net, where individualism and freedom of expression reign), but to keep the discussion on track and avoid wasting people's time. That makes moderated lists extremely valuable on the Net, as long as you can find ones that match your interests. There are hundreds. Thousands? Who knows? Enough that a list of mailing lists on the Net requires about 1.5 megabytes on a PC's hard drive. That's a lotta lists.

## Databases

Much of the available information on the Net is in the form of databases, like those on commercial services. One big difference is that virtually everything here is free, without the surcharges you find for some of the best databases on the commercial nets. That is likely to change, however, as the Internet gets more commercial, which is happening constantly and rapidly. Still, most of it will remain free because the information is either there for the benefit of profit-oriented companies, or those companies put it there because they want people to have it for some ulterior motive—usually to make themselves look good or to make you buy something.

Fortunately, the norm on the Net is to sell things indirectly, by providing useful or interesting information rather than hard hype. Whether it's census data or the latest mail-order catalog, you can often get it by FTP, which stands for *file transfer protocol.* It means that you can tell your computer to go search other computers, find what you want, and bring it back to you. Indisputably convenient.

## Telnetting

One major advantage of the Net is that you can *telnet*—telecommute via the network to other sites and read or download whatever you find there (as long as it's available to the public). Telnetting is like what the Star Trek crew does when they want to beam down to another planet. In this case, it puts you inside the remote computer so you can browse around the message boards and files there.

Remember that any online service is technically part of the Net. That means you can hop from one service to the other, as long as the service you want to go to doesn't require everyone to use their software to get in, as America Online, Prodigy, and Pipeline do. (On these services, people can get out and roam around; they just can't get in from somewhere else.) So you can be cruising around Delphi and switch to a Net newsgroup or jump over to CompuServe to check the latest buzz on a message board in a forum. However, on the fee-access, commercial services, you have to be a member of that service to get past the gateway. If this sounds way too futuristic, think of it this way: Logging into your local college library from home or into the LAN where you work while you're on a business trip is much the same as telnetting.

## World Wide Web

Also known as The Web, WWW, or occasionally W3, this is the other best feature of the Internet. In many cases, it's just the bare-bones, plain-text Internet all jazzed up with a graphical interface, so it has photos and graphics. Some Web sites, as they're called, even have sound and video. Now that most of the major commercial services and even some of the smaller ones are so graphics-heavy, in some ways it's not that surprising.

But another key feature of the Web—which the commercial services are adopting—is its underlying structure, called *hypertext links*. The net result is that you can click on a highlighted word on the screen or graphic and be instantly transported to whatever that word or graphic refers to, even if it's on the other side of the world in an entirely different domain. Think: *hyper* for fast + *links* to *text* = hypertext links.

For example, in looking for something for chapter 8 of this book, I started at the World Health Organization in Geneva, Switzerland. Something I found there pointed me to Stanford University in California, which showed me the way to something else at the University of Iowa, which then took me to Columbia University, just across the river from where I lived at the time in New York City. In less than 15 minutes, I was inside computers in those actual places, so to speak, and had traveled from Brooklyn Heights to Geneva to the West Coast to the Midwest, and ended up back in New York City. I would have gone to the Columbia site first had I known about it, of course. But I wouldn't have had such a grand tour, nor had this tale to tell you. And to think that Phileas Fogg thought it was a big deal to go around the world in 80 days.

– 33 –

That brings us to the end of our brief tour. Somewhat simplistic, perhaps, but there you have the outline of what it's like online, from the perspective of structure and function. Ah, I hear what you're thinking: "Big deal." What you really want to know, I'll bet, is what's so great about it that you should bother. Fair enough. In the next section and the next chapter, let's focus on what's there to make it worthwhile, with a bit about how to get online.

# How and whether to get on the Internet

There are many books written about nothing but the Internet, for good reason. It takes many books to explain how to navigate it and what's there. Before you hook up to the Net, it helps considerably to read one or more of them (see the Bibliography for some of the better ones). All the hows, whys and wherefores are beyond the scope of this book, but let's go over the most common questions and general guidelines.

The first question people usually ask about the Internet is: "Is it worth it?" Well, sorry to hedge, but the answer is: "Yes and no." Yes, because it's fascinating and an indispensable tool for finding certain kinds of information—the more esoteric or specific that information, the better—and for finding some kinds of experts. Almost any kind of either, in fact. No, because you can spend hours, days even, wandering around, feeling all the while like you're in another dimension (which you are, in a way). Then once you readjust to reality, you wonder, "So who needs all that information anyway?"

Well, we all need some of that information once in a while and some of us need a lot of it often. For students, teachers, writers, businesspeople and others, the Internet is a definite advantage. Increasingly, not being on the Net is a real disadvantage and undermines your ability to keep up with knowledge and news or remain competitive. But the most compelling reasons to be on the Internet are the myriad ways in which it widens your world and broadens your mind and feeds your soul. Oh, and it can also be great fun. So my short-form answer to the question "Is it worth it?" is yes.

Even if you decide to boycott or protest the whole thing, whatever your reason, you need to understand it, first-hand. You owe it to yourself to try out at least two different services, preferably quite different ones, for at least 30 days each. Then decide for yourself. That way, your netfreak friends and relatives can no longer accuse you of being just obstinate, technophobic, or a troglodyte. You'll have acted rather than reacted, and made an informed decision. It's your choice whether to stick with it or not. Depending on your needs and interests, you might do just as well sticking to the commercial services. Regardless, I guarantee that you won't grasp much of what the Internet is really like just by reading stories in newspapers or magazines. Or books, for that matter.

Just promise one thing, please—well, two: participate in discussions and explore widely. Otherwise, you might as well stick to photos in books and magazines.

The second most-commonly asked question is: "How do I get on the Internet?" That answer is much easier. It's far less frustrating, less intimidating, and smarter to try it first through the gateway access offered by one of the major commercial networks. That way you can browse around, get an idea of what's there, and decide whether you want to go further. If, after you explore the Net a bit, you think you want to subscribe to only one or two newsgroups or mailing lists or search for information once in a while, you'll probably be better off doing that through your primary network.

However, if you're instantly enthralled and think you'll use the Internet as much as five hours a month (it's easy to spend that much time roaming the Net in just one afternoon), then the best bet is to sign up for a separate service, one specifically for the Internet, known as an *Internet service provider*. That way you won't be paying the higher prices for connection through another service. If you have a local freenet, that's worth considering for basic access. Otherwise, even if your local BBS touts Internet access, you'll either have to pay extra for it or you won't get the Web's graphics, photos, and sound. Don't fall for the argument that you can use the Web browsers that the commercial services have or will have. Even if a few hours of your online time every month are covered by your basic fee, you're likely to exceed that when wandering the Net and run up a bill. The only way to really learn what's online is to spend a lot of hours exploring at first so you can pick and choose only what's worth it to you. To do that, you can't watch the clock and the meter.

Nonetheless, it's a good idea to keep your account on the commercial service too, at least for a while. You don't need to have both a commercial network account and one with a straight Internet provider, of course, but about 40 percent of people who use online services do, because each option offers distinct advantages.

The only way to decide what's valuable to you is to explore each of them during the free trial period they all offer. If there's something you really like or need on one service, but not much else there appeals to you, it still might be worth paying the basic rate to use that service for just one thing. Carola Draxler, whom you'll hear more about in chapter 8, keeps her America Online account solely so she can attend the Sunday-night fibromyalgia support group, for instance. You might want access to the kids' services on Prodigy or want to belong to Women's Wire because of its special focus and atmosphere. The commercial nets are also easy to get around and often provide a stronger sense of community than the Net, which is loosely structured, vast, and sprawling.

# Summary recommendation

Try the commercial services first, including whatever Internet-access services they provide, because they're easier and it's a good way to get oriented, no matter where you end up spending the most time. Then decide what you like best and need most. If you expect to be on the Internet more than five hours a month, use a separate Internet service provider for that, even if you keep your commercial net account. And don't discount the value of the many local and specialized BBSs and freenets, particularly if they're free or low-cost and a local call away. The next chapter describes a few of those, and gives you a sense of the atmosphere, benefits, and drawbacks of each of the main services covered throughout the rest of this book. For more information on what you need and how to get online, see appendices A and B.

# CHAPTER 3

# An overview of services in this book

**A**sk five people and you'll get at least four conflicting opinions about what each of the leading online services is really like. Each has pluses and minuses and each has its own character. The content—or what's there—varies too, but more in depth than in breadth. Free trial periods and low monthly rates aren't the only things that get people to try a service, nor stay once they do. As many as 85 percent of the people who try a service don't stay. Of those who do, it's generally the combination of the particular character or ambiance and the content, plus the friends they make there, that keeps people paying the monthly tab. Although the evaluations that follow reflect some degree of consensus among those who've used these services, they're inevitably colored by my own opinions. So consider this a guide, not gospel.

To balance my own subjective impressions with at least a modicum of objective criteria, I asked a small group of women to rate each of the services on a four-point scale, with one as low and four as high, on the following 15 factors: overall substantiveness of content, breadth of services, real-name policy in public forums, services for kids under age 16, services for professionals and business-people, services specifically for women, comfort factor for women, social-consciousness forums (education, environment, politics), home-related services, leisure-pursuit services, time-saving features and services, ease of use, price (considering the basic fee and rate for additional hours or surcharges), quality, and reasonable cost of Internet access.

The respondents were all women, at my request, and self-selected from messages I posted online. All had used more than one service, but all the services they used were ones with graphical interfaces. I knew none of them personally. Interchange and the Microsoft Network are not rated because neither was officially open for business when I conducted this survey in early 1995. Echo and the WELL aren't included either, because I didn't post the call for responses there; I rate them both at roughly 2.5. Despite the fact that this is an unabashedly unscientific survey, the results are probably not far off from what you'd get from the experts:

| Service | Rating (1–4) |
|---|---|
| America Online | 2.46 |
| CompuServe | 3.29 |

| | |
|---|---|
| Delphi | 2.07 |
| eWorld | 2.29 |
| GEnie | 2.50 |
| Internet | 2.88 |
| Prodigy | 2.25 |
| Women's Wire | 2.75 |

Because the price structures are often confusing and easily misinterpreted, I also compared what it would cost to spend 20 hours a month on each of these services, half sending and receiving e-mail and half in the forums, based on evening rates of 9,600 bps or faster, if that made a difference:

| Service | Price per month |
|---|---|
| America Online | $54.20 |
| CompuServe | $57.95 |
| Delphi | $57.00* |
| Echo | $19.95 |
| eWorld | $56.15 |
| GEnie | $40.95 |
| Internet | $17.50 |
| Prodigy | $54.20 |
| The WELL | $55.00 |
| Women's Wire | $77.10 |

* Estimated because of pending price change

Bear in mind that *Jupiter Communications' 1995 Consumer Online Services Report* defines "medium users" as those people who spend an average of 8 to 15 hours a month online, and "light users" (60 percent of the total) as those who stay within the monthly minimum and incur no charges beyond their basic monthly fees. Therefore, most people would not spend this much, although some spend far more.

Also, Internet access is a significant wildcard here. For these rates you could get anything from nothing but Internet e-mail all the way to full Web access. But that's the way it is in real life. The Internet rate of $17.50 is based on *PC Magazine*'s annual online networks report in February 1995 for the average cost of direct Internet access, but this wouldn't likely include a SLIP or PPP connection for full Web access. In all likelihood, because of the perpetual price wars some of the rates will have changed, but the information should give you a way to see through some of the advertising hype.

What will surprise most people is that CompuServe's fees come out comparable, despite the constant complaints about their rates no matter how often they lower them. The reason is that e-mail is included in the basic monthly fee and, if you choose their standard pricing plan, you can send 90 messages to your friends and relatives; whereas on the other services you start paying after the allotment of free hours each month. Thus "free" can turn out to be more expensive. One more reminder: You'll save time, thus money, on the services where you can use offline-reader software and not be forced to access the forums online, with the cash register clanging in the background. See appendix A for details.

The bottom line is that if people find an online service mildly amusing, but no more, they'll pay the price of going out to a movie instead. If it's worth more to them, they'll pay more for it, just like anything else. Yet there's not necessarily a correlation between value received and price paid. Numbers certainly don't give a clear picture, and subjective factors are just as important as objective ones, so let's look at a snapshot of each service.

# America Online

*Pluses* Attractive, easy-to-use interface, classes, parental discretion features, magazines, smooth introduction to the Internet.

*Minuses* Hype about People Connection chat rooms has created an adolescent atmosphere in some places that can be inhospitable for women. Even if nothing happens, you feel wary, because you hear so much, so often about "harassment." (For the record, I use my last name on AOL, only because no more letters will fit in the allotted space, and I've encountered no problem.)

*Best use* Leisure pursuits, but potentially good for courses and personal finance.

AOL, as it's called among the cognoscenti, has a rather rah-rah, youthful aura, which is reflected in the half-dozen or so graphics that pop up on nearly every screen (see Fig. 3-1). It's perceived as a service for people under age 40—the average age is 32, says Jupiter—even though it's the home of one of the best services for people over 55, SeniorNet. It's been around longer than people think. AOL began as a Mac-only service, and nearly withered on the vine until they introduced a graphical interface for Windows users. While it was still much smaller than Prodigy and CompuServe and the giants were still charging new members for their software, AOL started giving their software away at corner newsstands with computer magazines and mailing it to anyone who ever said the word *computer*, it seems. Because their software has been the most attractive and easiest to use and because they've promoted the service more aggressively than any of the others in the last couple of years, it has grown fast, and now claims two million members.

**3-1** *America Online's log-on screen.*

But until early '95, people commonly complained that, as Gertrude Stein might say, there was no *there* there. "Content poor," the new media moguls called it. Meaning, in both cases, not much substance. The buzz now is that they've reached critical mass and are changing. Actually, the structure for a great service has been there all along, and I'll hit some of those high points throughout the book.

AOL, unfortunately, got sidetracked by hyping their numerous People Connection chat rooms, with names like Flirts' Nook and Romance Connection. Like other services, that and the game-playing areas are where they make the most money, plus some of the honchos seem to think that the key to keeping people online is creating a sense of community (partly true), and that the way to do it is to get them to type at each other in the chat rooms (not true). This has only reinforced their lah-dee-dah image and gained them a reputation among women as a place where some bozo is likely to make sexual advances if you have a female name. To make matters worse, AOL's sign-up procedure encourages fake names, and their system allows up to five different names per account. That's convenient to give all members in a household their own online identities, but it invariably leads to abuses of the privilege, thus they flunk the real-name-policy test.

Too bad if they think razzle-dazzle will work long-term. What's more likely to happen once the hoopla over the information superhighway dies down and people think of it as just another part of life or become disenchanted is that people will try AOL and discover only the superficial things about it, and then either log off permanently after their free trial periods or log on only occasionally. Most of those who are disillusioned by their first encounter with the online world, based on AOL or any other service for that matter, will judge all other services by that one experience.

AOL is organized as though things were tagged on as afterthoughts, so you have to search through two or three layers of menus to find some of the more interesting places. Among the three best things are the Interactive Education Service, the Better Health and Medical Forum, and the Issues in Mental Health Forum. Modems and online communication are going to have a tremendous impact on education, and AOL is the only service other than GEnie, so far, that's doing much in the way of offering classes. For sheer quantity, they also have the most online support groups, and reports about their quality, particularly those set up by the forum sysops, are positive. They've also done a lot in the last year to strengthen their personal finance areas. CompuServe, Prodigy, and sometimes GEnie are still the most popular for money management, but AOL is gaining ground. Also see chapter 20 for information on their new women's forum, Wired Women.

They're big on "IMs" here, or instant messages, meaning you're going about your business and some stranger will interrupt you to strike up a real-time conversation. Those dialogues are as meaningful as conversations at crosswalks on the street, and can sometimes start with blatant sexual propositions. For that reason, you'll hear a small contingent of women constantly grousing about violations of TOS, or terms of service, which has the ring of, "If you don't play my way, I'm going to take my ball and go home" or "Daddy, make him stop!" The thing to do is avoid such people and ignore these types of messages.

Overall, AOL is a well-done, easy-to-use service that's getting better as it grows—or grows up. But part of what makes it easy to use is also supremely irritating. It's nearly impossible to log on without having to wait while the system downloads graphics into your computer, while you sit and twiddle your thumbs (and pay), whether you want the graphics or not. The other real drawback is that the messages aren't threaded and ones addressed to you aren't flagged, so you have to jump all over the board looking for replies. Because you can use only their software, you're a captive audience, and you have to sit in your chair and click away at stuff on the screen. It's enough to discourage many of us who are long since addicted to software that does everything while we're out getting coffee, and are used to forums where we can easily follow one discussion from beginning to end. It's just too time-consuming.

# CompuServe Information Service

*Pluses*  Most depth and breadth of the commercial services, especially for business and career interests. Best research databases. Forums have the most substantive discussions and most well-stocked libraries. Several offline-reader software programs are available. Widest international reach.

*Minuses*  Some of the most useful services are surcharged, so it can cost more.

*Best use*  Business and professional interests, news, research, some leisure pursuits.

CompuServe (see Fig. 3-2) is the oldest, largest, and most serious of the commercial services. It has the most business-oriented forums and services and the widest range of software and hardware tech support by far. Most vendors maintain at least a section, if not a whole forum, on CompuServe, although getting a timely answer from some of them in the midst of a computer crisis can be as frustrating as trying to reach tech support by phone. CompuServe's research services are unparalleled, except on the Internet, and you can often find the same information faster and more easily here, even though you'll have to pay by the minute. If you learn tricks and shortcuts for things like ZiffNet's magazine, health, business, and computer databases (see chapter 18, *Online research basics*), and the Knowledge Index, the trade-off in less time wasted and less frustration is usually well worth the price.

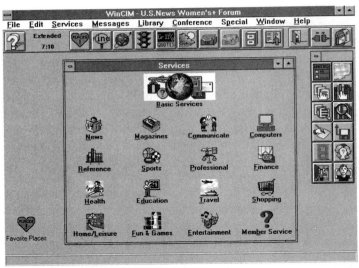

**3-2** *CompuServe's opening screen.*

Because CompuServe users paid more to be there than on the other services until recently, it attracted people willing to pay to be among other people for a purpose, not just for chitchat. Thus there are more experts and people with solid expertise here than any place except Internet mailing lists. The two other largest services, America Online and Prodigy, have positioned themselves as "consumer services," meaning after-hours use, thus the business folks tend to be on CompuServe instead. As more people get online and shop services and the rates become more comparable, there's a noticeable influx of people used to different standards of online behavior. But the culture on CompuServe is a bit more formal and dignified. The members like it that way and react coldly to those who violate the norms, so newcomers either adapt or leave.

This isn't to say that people on CompuServe aren't friendly, because they're just as helpful and cordial as on other services, and many of the forums have periodic offline get-togethers in various cities. Nor is it that nobody's here just for fun. There are forums for almost any hobby or sport; a games area; excellent places for music and movies, both for artists and fans; and a terrific travel forum, to cite just a few.

The sysops have a much stronger presence here than on most other services, and there are several in each forum, so that tends to keep things on a higher plane and results in prompt responses to questions. CompuServe has a large chat area called CB Simulator that has its devotées, but for most users it's out of sight and out of mind. CIS doesn't promote it, as AOL and Prodigy do their chat areas. They also have a real-name policy in all forums but the self-help ones, which goes a long way in creating a polite and more mature atmosphere. At an average age of 42, according to Jupiter, CIS users are also a tad older, and more of them are male than on the other commercial services. The latest independent audit says only 12 percent of the members are women, although CompuServe says their own survey shows 17 percent, because many women use their husbands' accounts. As one who uses the service daily and has for a few years, that figure sounds about right, whereas I'm skeptical of the claims on a couple of the other services.

The CompuServe Information Manager software is designed more as an orientation and browsing program rather than as the best choice for more experienced users. To my mind, it's as easy to use as AOL's software, although some would disagree, but I think that's only because it isn't as graphics-intensive. Unlike the other services, you can turn some of the graphics off, such as forum logos, which saves time online and space on your hard disk drive. There are also more third-party software programs available for CompuServe than any other service, most of which are excellent, and all of which automate more of the commands than the CIM software does. CompuServe also offers an offline reader program called Navigator for intermediate-level users. Any of these programs save you time and money (see appendix A for details).

The Executive News Service, which is an electronic clipping service for the newswires, and many other useful services that other services charge for are included in your basic monthly fee (although CompuServe charges for some things others don't, as well). All in all, CompuServe is hard to beat for value, usefulness, and substance.

# Delphi Internet Services

*Pluses* Low cost, excellent Internet access, ability to browse Internet newsgroups.

*Minuses* Relatively few core services or forums and an antiquated interface that's no fun to use, although they're about to incorporate Netscape, the most popular Web browser for the Internet, and add graphics and several services.

*Best use* E-mail and Internet access, unless things change radically with their new plans to reposition and revive the network, which is likely.

The name kinda says it all here: Delphi *Internet* Services (see Fig. 3-3). They're the only commercial service that has offered true full Internet access until rather recently, and because they've licensed use of Netscape, currently the most popular and most powerful software for browsing the World Wide Web and for other Internet functions, it looks like that's going to continue to be their thrust. However, they're in the process of a major overhaul and weren't divulging details when this book went to press, so it's too soon to tell where Delphi will end up in the lineup.

Aside from their Internet access, most of the good stuff on Delphi is in the few hundred custom forums, all of which were created by members. There's a forum for women, and a few other special-interest forums that have no counterparts on other services, but I can't say that it's worth joining Delphi just for those because it's so limited otherwise. As other services have bypassed Delphi

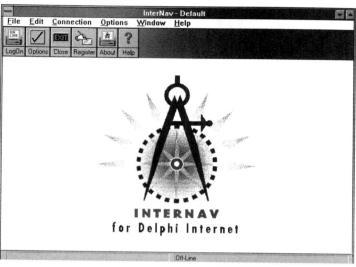

**3-3** *Delphi's log-on screen.*

in ease of use and breadth of services, the members who remain there are die-hard loyalists who like it because it's smaller and not overrun by window shoppers. They also like it because it costs less, and are willing to put up with typing in commands to navigate for those reasons. Delphi introduced InterNav, a Windows program, in late 1994. About the only thing it automates is the log-on process, and the only way it helps once you're online is that you can click on a word or a number to move ahead. It's very clumsy, not much of an improvement over plain manual commands, and it bombed. So they're revamping, and will introduce graphical software in 1995 for the first time.

The other good thing about Delphi is that it's one of the three services—the others being CompuServe and GEnie—where you aren't restricted to using whatever software they provide, thus you don't have to let the service control how much you have to remain online, paying all the while, to accomplish what you want. Independent developers have created script-based programs that enable you to get and send mail and get new messages from forums, all automatically, and all in a few minutes. D-Lite for DOS, its counterpart Win-D for Windows, and Rainbow are the most popular among those. You can download them online, try them for 30 days, and keep using them if you like them for a registration fee of typically less than 50 bucks.

Unless they surprise us with a whole new concept and great new software, the only strong reason to use Delphi is to avoid wrestling with installing Netscape and for their reasonably priced Internet access, with their other services as ancillary bonuses. Even then, you might be better off signing up with a local Internet access provider that also has a few forums and entertainment listings that focus on things of local interest.

# Echo (East Coast Hang-Out)

*Pluses* Erudite, New York City-oriented chat; emphasis on the arts; cordial atmosphere for women; strong sense of community; low cost for core service.

*Minuses* Text-based interface for the core service, with cumbersome commands compared with a graphical user interface, although not difficult if you use it often enough to remember the commands.

*Best use* As a hangout to converse with rather sophisticated people in a well-run online haven devoid of sleaze. Echo might branch out to other cities, but each affiliate would focus on the local city or region.

To be fair, I must say from the outset that I wasn't able to spend much time on Echo in preparation for this book, compared with the bigger services. I've been a member for some time, however, and used to log on regularly when I was new to New York City, because it was the best way for a stranger in a strange land to get oriented. Want to know where to get the best pizza in New York City, or find out which are the best bookstores, restaurants, even doctors and dentists? Echoids will know, and will go out of their way to tell you everything else you need to know to survive in New York, from negotiating a lease to finding a job. Because the arts and publishing are major industries in New York, those are major topics of conversation here.

Because Echo keeps the rates low, many of their members log on daily and make the rounds. Echoids, as the members call themselves, get together regularly at a café in Manhattan and at private parties, and Echo sponsors events such as their Read-Only series, readings by authors who are Echo members, and their Virtual Culture series, which is about the arts and electronic media. All of those extra efforts, combined, have created a strong sense of community. Echo also offers either plain Internet access or full-scale SLIP access so you can use all features of the Web. Either is available without joining Echo itself, but Echo members get discounts to Internet services. They also now have a Web site for visitors, at http://www.echonyc.com.

At first glance, someone might wonder why I wanted to include Echo and the WELL in this book, in the same context as America Online, CompuServe, and Prodigy. For one thing, they're both major services in their way. For another, many prefer them to the more mainstream, bells-and-whistles networks, or use them as their central place to socialize but use the bigger services for other reasons.

# eWorld

*Pluses* Attractive, easy-to-use interface with a slight emphasis on the arts and education. Commendable real-name policy, and a conscientious effort to create and maintain an atmosphere where women and children feel welcome and comfortable. Interesting services for kids too.

*Minuses* For Macintosh users only, although Windows software and full Internet access for all, including the Web, are due in 1995. Not enough depth and breadth in services, and too many forums sponsored for commercial pur- poses with very little participation.

*Best use* For tech support and Mac-related information, perhaps for parents to be involved with kids who use Macs at school, for musicians to exchange tips about creating electronic music, or for filmmakers to discuss how to use Macs in film editing. However, all that can also be done on some of the other services.

eWorld, shown in Fig. 3-4, is still relatively new and small, although they'll have an influx of about 60,000 people this year when Apple Computer merges AppleLink into eWorld, plus many more when they introduce software for Windows users. During the few weeks I lurked around, only people who use Macs could use eWorld.

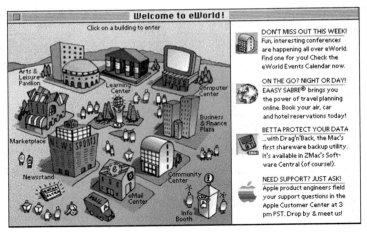

**3-4** *eWorld's main screen.*

In a discussion started by the general manager about why only 20 percent of eWorld members were women at the time (February 1995), several women said they like eWorld because it's small and friendly, and they dread the invasion from what some actually called "the dark side," meaning PC users. The attitude toward people who use a different kind of computer system was amusing and amazing, and made eWorld seem like a benign cult. People on networks dominated by PC users still sometimes make disparaging remarks about Mac users, but they're in jest, not serious. These folks were serious!

Because Apple is based in the Silicon Valley area of California, there's a West Coast slant to many areas, such as nightclub listings, promotion for bands (although there's one band online from Australia, and possibly other countries too), and environment-related services. There are early signs of a trend toward locally and regionally focused areas on even the largest commercial networks, and the Internet has had that all along. Because eWorld already has some seeds planted to grow in that direction, that could become one of its assets. They seem to be playing to their strengths somewhat, with an emphasis on the arts—particularly theater and music—as well as a few imaginative features for kids and teenagers. But the short of it is that there's simply not much happening here yet.

As it stands now, the main motivation for joining eWorld would have to be either for something very specific and compelling that no one else has—and nothing comes instantly to mind—or because you prefer a small, friendly service yet want the easy-to-use graphical interface that seldom comes with a local BBS. If they're going to turn it into something beyond that they'll really have to hustle, or else carve out a niche that sets them apart from the majors.

# GEnie

*Pluses* Solid content in most discussion forums, good research services, sense of community.

*Minuses* Although it's ranked as one of the majors, its forums aren't as extensive as the three leading services. Their software has been a major handicap, and the vote wasn't in yet on the new software when this book went to press.

*Best use* Research for small businesses, personal financial management, classes, hobbies, and games.

It doesn't feel good to say bad things about GEnie (see Fig. 3-5). It used to be one of my favorite services because its content wasn't as airheaded as some of the oth-

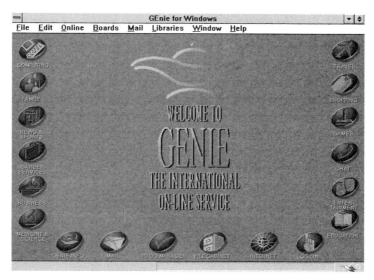

**3-5** *GEnie's main screen.*

ers, and because there's something about it that makes it almost endearing. You want to like it. But General Electric has neglected it, and let it lose most of the foothold it once had. They raised a lot of excitement among their 400,000 fans in anticipation of the Windows and Mac graphic interfaces they introduced in 1994, then both programs turned out to be very disappointing. Many folks quit waiting for things to change, and left. Those who stayed are a loyal crew, and GEnie is now trying to reinvent itself. It introduced yet another whole new look and new graphical-interface software in mid-1995, and plans full Internet accesss—finally, long after everyone else. None of that was ready when this book went to press, so it's too soon to say if they'll live up to their name and work magic.

GEnie's policies about real names and standards of decorum are excellent and the sysops have a strong presence, all of which help make it a pleasant place to be. The people participating in discussions are generally knowledgeable and helpful, and GEnie was once one of the most popular places for personal finance advice, and still is a hangout for many people who've been GEnie members for years. Their research services are second only to CompuServe's, although they cost extra in both places, and the games are popular. But by not keeping pace with the movement toward easy-to-use software and graphics-based commands, they lost their momentum. To regain some of it, they've lowered the prices. Along with the new software, this might give people the impetus to try GEnie, and help put them back on track.

They seem to be aiming for the younger crowd, which is apparent even in their logo, although when I've been online there I've never had a sense that I was surrounded by Generation X rather than my over-40 peers. So, having seen only a glimpse of what's coming, the best I can say is keep the faith. GEnie has been a good service in the past, and it's definitely worth giving another chance if you dismissed it before.

# Interchange Online Network

*Pluses* Excellent, elegant interface that's easy to use, well thought-out, and appealing, with features unavailable on other services. Signs of good content to come.

***Minuses*** Too soon to tell.

***Best use*** Too soon to tell.

Interchange, shown in Fig. 3-6, was still in the beta, or prerelease, stage when I researched this book, and was gradually growing. But because it wasn't completely set up nor even open to the public yet—although it should be by the time you read this—it would be unfair and misleading to try to assess much about it here.

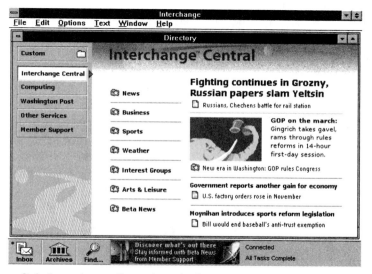

**3-6** *Interchange Central on the Interchange Online Network.*

One of the distinctive things about Interchange is its software. It's available only for Windows, but there's a Mac version in the works. When they first designed it, it was so different that it was almost futuristic compared with what others had. But by the time they were through the testing stage, the World Wide Web had become "the thing" and several other services were about to introduce redesigned software, all of them incorporating one of the key things that set Interchange apart: hypertext links. This means that when you read a message, there's an icon or graphic you can click on that will instantly take you to more information on that subject, and even to other places on the system that are related. It sometimes makes you feel you're on a ride at Disneyland, but it's cool and convenient, nonetheless. In the forums, you can read topic headers and part of the related message at a glance, which makes it easier and faster to choose what to read or what to skip. You can also tag threads that you want to follow, which links them to your mailbox so you can keep track of new messages in those discussions for as long as you want, and read them offline. This is a great time- and money-saving feature.

Interchange and Prodigy are going to end up looking the most alike, from the previews I've seen of where they're both headed. Interchange looks prettier and more artsy and elegant, and the character and content of the two services will make them different enough to attract different kinds of users. Interchange was conceived as a tool for publishers, so they've already made deals with the computer magazines formerly owned by Ziff-Davis and are likely to go after more magazines, which has been AOL's turf and to some extent CompuServe's. Prodigy has aimed for newspapers instead, and has lured the *L.A. Times*' TimesLink, among others, one of the best incarnations of any newspaper online. But Interchange hosts the *Washington Post*'s online version, Digital Ink. Both cost extra after free trial periods, whereas the *New York Times*' version on America Online, called @times, is free, but has selected stories from the current day only. IChange, as it's already known, will also have the usual after-hours services about home maintenance, hobbies, sports, and such. Finally, Interchange and the Microsoft Network are both likely to give CompuServe competition for its long-held dominion over the computer hardware and software online tech-support realm.

Although it was started by Ziff-Davis, Interchange is now owned by AT&T, and you can bet they intend to build it into something to compete with the other major companies. So add this one to your must-try list, but bear in mind that it will be like visiting a new town erected in the desert at first, because it takes time for a new service to develop and grow.

# Microsoft Network

*Pluses* Ease of use, possibly quality of content, probably basic price, although most areas will cost extra as they do on many other services.

*Minuses* Only for Windows, and it's up against serious competitors who are serious about doing whatever they can to keep Microsoft from having the dominance that controlling the Windows operating system automatically gives them. Major disk-space hog. Too soon to tell, otherwise.

*Best use* Too soon to tell.

The Microsoft Network (see Fig. 3-7) was also still in beta stage when I tried it, and not due to debut until August of 1995 or later. Plus, they insisted on a nondisclosure agreement, so there's not much to say. It will be easy to use, although it's part of the next generation of Windows, which is a major change from the current incarnation of Windows. Because of this, there are inevitable problems that could take several months or up to a year to fully fix, so before you convert your whole operating system and a lot of the software that depends on it, wait until those glitches are under control. If you're on any other online service, you'll hear more than you ever wanted about it through the grapevine, so judge by that.

– 55 –

There will be two women's forums on MS Net, along with all the usual things you might expect for leisure and business use. Because Microsoft is offering

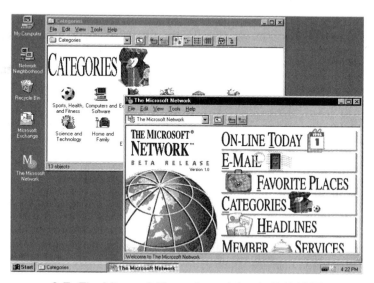

**3-7** *The Microsoft Network, to debut in Fall 1995.*

content providers—the people who create and run forums—a chance to keep more of the income than other services allow, it's likely to attract a slew of new services, and some who have been online elsewhere might even switch shops. That could have its downside too, in a number of ways.

If you use Windows, there are enough advantages to the new version that it will be worth upgrading after the bug phase. One of the options during installation is to sign up for MS Net, and few people will be able to resist that easy opportunity—which is just what they're counting on, of course.

# Prodigy

*Pluses* Excellent services for kids, Women's Leadership Connection, personal financial management services, Newsweek Interactive, *L.A. Times'* TimesLink, and the best parental discretion features available.

*Minuses* No libraries for forums, variable quality of content.

*Best use* Leisure pursuits, family use, shopping, personal finance.

Serious competition has been good for Prodigy (see Fig. 3-8), because it has forced them to beef up their content. They've always billed it as a consumer service, with heavy emphasis on the irritating ads that pop up at the bottom of the screen with every move you make. What I've never understood is why they're the only service that couldn't turn a profit for a decade, yet they were the only one that sold ad space. The answer is probably partly because they invested a great deal in creating the first graphical user interface and, even though those are now the norm, it took them a long time to break even on that initial outlay. Ironically, their interface became clunky and garish in the interim and they had to redesign it all.

You can enter separate names with separate passwords for up to six members of your household, yet control what services each has access to, with different services for each one. Thus you, as the master account holder, can allow your teenager into different places than your third-grader, and prevent both of them from going into the Plus surcharged services if you want (see chapter 5, *For*

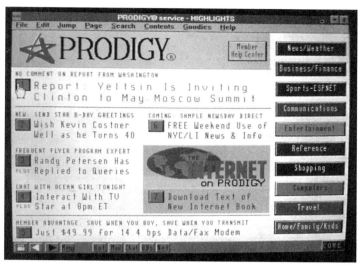

**3-8** *Prodigy's main screen.*

*parents, about kids*, for how to do this). People tend to use real names on Prodigy, or at least not cryptic or suggestive ones as they do on AOL, which makes this similar setup a benefit rather than a liability.

America Online is beginning to encroach on Prodigy's territory as the leader in services for children, as they are in the personal finance area. Although GEnie and CompuServe also have areas for kids, they don't come close to some of the things Prodigy offers, thanks to sponsors such as Nova and NASA.

Prodigy still has an image as a service for dilettantes—people who are just playing online or who maintain the account for their kids. That perception is mainly a holdover, and is no longer as valid as it was when the quality of its content was more shallow and uneven. There are now enough high points to again make it a major contender. That's always a matter of personal perspective, so what I think is great you might think is ho-hum. I mention various sections throughout this book, so judge for yourself whether you think it has what you want.

The Women's Leadership Connection is still young, but has the potential to become a truly good service. You have to request admission even for a free trial, then it costs an extra $7.50 per month on top of your Prodigy fees if you stay after the first 30 days. The good thing about that is that it might keep the flamers and those just cruising for cybersex at bay. (See chapter 20, *Women belong online*, for more.)

Prodigy charges for most things in some way, and the constant "buy this, buy this!" is hard to ignore. For that matter, you get bombarded by ads on the Web too, but it's more information- and service-oriented, so a bit easier to take. When I'm on Prodigy, I always feel like someone's constantly shouting at me. Others learn to tune it out more easily than I, obviously, or there would have been a mass exodus long ago. Not that there's anything really wrong about having ads. We're used to them with newspapers and TV and radio shows, and we might as well get used to them online, because more and more services will have to find ways to make money from the people who spend only basic-rate time online.

Yes, there are things here of value now, even for grown-ups. For children, Prodigy merits being at the top of your try-out list.

# The WELL
# (Whole Earth 'Lectronic Link)

*Pluses*  Witty, interesting, erudite conversations. Social consciousness.

*Minuses*  Command-line, straight-text interface that's difficult for the uninitiated to use, but by the fall of 1995 they'll have a graphic interface.

*Best use*  Good conversation.

I can't claim to be thoroughly familiar with the WELL because I haven't spent enough time on it, and one of the reasons I haven't spent enough time on it is that it's so difficult to use when you're used to doing everything with graphics-based software and point-and-click commands on other services. It takes longer than usual to get oriented and, once you are, you simply forget the commands unless you log on frequently, so it's like starting all over again if you stay away for very long. They now have a menu option, which will help. I'm not computer illiterate, you understand; it's just a matter of time and patience, and I have little of either. Because time is a major roadblock to using online services for most women, you have to be highly motivated to become a WELLbeing.

The reward for your perseverance is that this is one of the most interesting networks, solely because of what people say and how they say it. Sometimes it seems like a contest among the self-consciously hip to see who can be the most witty or clever, and it can even come across as an enclave of smug, self-satisfied liberals. But many of them are very well-informed, and they think through issues and carry on rousing debates. So it's entertaining and has more personality, overall, than any other service I've ever seen.

A few years ago, in a special issue on what they claimed was a movement or revival of salons across America, *Utne Reader* called the WELL an electronic salon. Probably because of the connotation of salons as a pastime of intellectuals, they liked that a lot, and still use it in their publicity. The salon analogy is a bit overstated, however, because a lot of the conversation here is simply inane chatter, and the Literary Forum on CompuServe could just as easily qualify as an online salon. Still, there's some truth to it, and the WELL is fun and a good place to flex your mind's muscles.

The WELL has long had regular off-line get-togethers, and it has a very strong sense of community. Alas, most gatherings are held in the San Francisco area, because the WELL's home is in Sausalito. Like Echo, they now have a Web site for visitors (http://www.well.com).

# Women's Wire

*Pluses* Cordial atmosphere for women, intelligent and thoughtfully developed content, easy to use. Browsable selected Internet newsgroups, potential for making regional connections.

***Minuses*** High cost, no libraries for all sections, no easy address-book feature in the software. Still too small for a wide cross-section of knowledgeable people and content, and too much of a Bay Area slant, although that's changing.

***Best use*** As a place to learn, confide, ask questions you feel intimidated about asking elsewhere, and as a place where you don't have to be *on* all the time as you sometimes feel you must with services where men outnumber women.

Frankly, I didn't think I'd like Women's Wire (see Fig. 3-9). That's because I've always felt that the place to be is on the same playing field as all the other players, and because so many of the sections for women elsewhere online had invariably degenerated into nothing but trivial chat, or even cloying sweetness or whining. I felt those fed every negative stereotype about women, and I certainly wasn't willing to pay by the minute to be there. So I was afraid that Women's Wire was going to be all about natural childbirth, recipes, and rock gardens. Wrong. In fact, I don't recall a thing on Wire about the latest lipstick or decorating or cooking light. You're more likely to find something on nutrition or an alert about toxic effects of some dye used in lipstick. If anything, it's almost too serious. But I happen to like serious, and it's past time someone

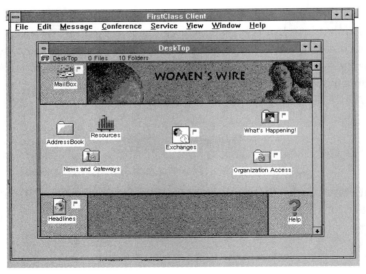

**3-9** *Main screen of Women's Wire.*

realized that women are interested in something besides domestic concerns. Women's Wire does.

Oh, you can find most of the things you'd expect here. *New Woman* magazine sponsors a relationships section, for instance, and their outlook gets pretty gushy and ga-ga. But the best thing about Women's Wire is that, even though the service is still young (it began in January 1994) and still small (1,400 members a year later), they've somehow managed to create an ambience that is undeniably special. It feels cozy and safe. Safe not in the sense that you aren't likely to have to fend off sexual come-ons—although that's true because 90 percent of the members are women, and although there are about the same proportion of lesbians among them as you'd find anywhere in the real world, lesbians aren't known for making unwanted advances. But no, not just that kind of safe. It's more safe in the subtle way you feel at ease when you're among any gathering of women, because you know there are some things that you all understand and take as givens, and you don't feel like you have to be somebody you're not just to make sure you measure up to some arcane code. There's also a tacit understanding that women here will try to help each other further their careers and solve problems.

– 61 –

As this book goes to press, Women's Wire had just announced that it will take over the women's forum on CompuServe, started by *U.S. News & World Report*, and run a sister service on Microsoft Network in addition to their original independent service. That could be a great boost for them to get enough members for their own service to survive long-term. Yet it could be their undoing as an independent, not only because it would divert the energy they need to build the primary service, but because people on those mainstream services will judge Wire by what happens there, yet they'll have less control over who's there and what happens than on a far bigger, more diverse network.

Although some were skeptical about how long Wire would last, they now have an infusion of investment capital and a new president, are geared for a great growth surge, and hope to one day grow into their full name: Women's Worldwide Information and Resource Exchange. What will be will be. But I can't help feeling a bit about this service the way the people on eWorld feel about having discovered a small, special place that they want to protect from invasion by the masses. For more on Women's Wire and other places especially for women, see chapter 20.

# World Wide Web and the Internet

*Pluses*  Range of information and choices, relatively low cost.

*Minuses*  Range of information and choices.

*Best use*  Research, learning, fun, finding kindred spirits who share your interests.

Some pundits predict that the World Wide Web will draw far more people three years from now than all the commercial services combined, and that the commercial nets will end up like ghost towns. Not so. As many advantages as the Web has, it and all the rest of the Internet are overwhelming and will become only more so. Plus, the population on the Net is much more diverse and always will be, so people will still congregate on the commercial services, not only for the convenience and because it's easier to find and do things faster in most cases, but for the same reason people identify with neighborhoods when they live in big cities.

That said, most of us will be both places if we're online at all, thanks in part to the ease of using gateways through the mainstream nets, but also because the software, installation, and navigation (see Fig. 3-10) will continue to get easier, and the Net costs far less.

The Internet is so vast that it can seem infinite. You can fritter away hours, days, or weeks just exploring, and exploring is the best way to find what really appeals to you. I've concentrated more on The Web in this book as it will increasingly be the way people explore the Net, because it's so much easier to use. To find newsgroups, most people will first investigate them through the commercial service gateways, and from word of mouth. The value of the Net isn't so much its vastness but that, with persistence, you can find Web sites and mailing lists and newgroups that satisfy even the most esoteric interests.

There are few among us who have never felt they didn't fit and longed to find true kindred spirits or others who shared our particular passions. You're more likely to find those people and places anywhere online than in the more narrow

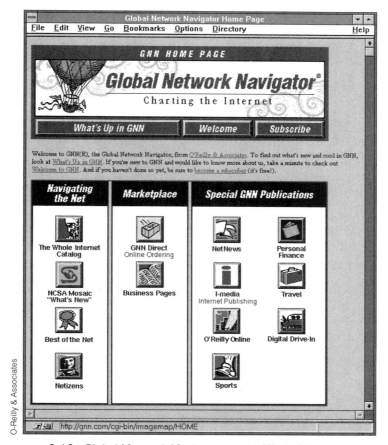

**3-10** *Global Network Navigator on the World Wide Web.*

confines of your own real-world community. And you're more likely to find them on the Internet, somewhere, than on the mainstream services. So, yes, to most of us it's worth the effort. You can still live a full and happy life if you're never on the Net. But you might wonder, once in a while, what you missed.

# CHAPTER 4

# Friends and family

When we first go online, most of us rationalize that we're doing it because we need it for work or for the kids or for any defensible reason other than what makes us stay, once we're there. The real reasons are that it's fun, and it's a great way to keep in touch with friends and family, and meet kindred spirits you probably would never cross paths with otherwise.

# The virtual community is real

"I hear lots of moaning and hand-wringing about how we, as a society, are losing our sense of community and becoming isolated," says Anna Conti, of San Francisco, "and how 'this computer stuff' is exacerbating the problem—how we're all becoming zombies who would 'rather interact with a computer than with real people.' Funny thing is, I never hear this from people who are online. I only hear it from people who go home every night to zone out in front of the TV, and wish there were more time to visit friends. . . . I have met more interesting, like-minded people who have become close friends in the two years I've been online than in the ten years before that.

– 66 –

"In fact," Anna continues, "the number-one best resource that I use for real-life problem solving is my online community. When I have a problem, I talk it over with them. I get real support in return, as well as practical advice. Recently, for example, my question was the classic one: What do I do with the rest of my life? I was burnt out on nursing, suffering a variety of physical ailments, trying to make up my mind about having surgery, generally feeling pretty stressed and depressed.

"I hashed it out over a period of about a month with my online buddies, occasionally face to face with a few of them. . . . I eventually started to get the idea that I'd like to revive my long-lost dream of being an artist. But I was afraid of starving to death. Then I heard from three artists in one week who all told me that they were poor a lot of the time, but it was worth it . . . . So I took the plunge, and handed in my resignation. No matter how it works out, I know where to go if I need advice, feedback, or just a sympathetic ear."

Anna's case for community is right on target. So much so that *virtual community* has become as common a term online as *virtual reality*. Just like any other community, to become a part of it and enjoy all of its benefits, you need to contribute and participate in it regularly. So if you simply browse around online or always lurk—meaning only read messages without replying to any—you'll never reap those benefits and you'll wonder, "What's the big deal?"

As Anna mentioned, people do frequently get together and become friends offline after getting to know one another online for a while, and many forums have picnics, dinners, or parties periodically in different cities. But even when relationships remain strictly online, sustained through e-mail and messages within public forums, the bonds can become strong. Online communities also develop their own traditions and customs, just as real-world ones do. In the Women's Issues Section of the Issues Forum on CompuServe, members give each other "virtual cinnamon rolls" as welcome gifts or to cheer a member up when she's feeling down. As you'll see in this excerpt from a thread in that section, they also often give each other far more:

```
Subj:  Our Virtual Community   Section:  Women's Issues Section
  To:  All                               Issues Forum
From:  Bev Sykes [Section Leader]        CompuServe

A community is a group of people who come together and make a place
where they can all share each other's lives. Sharing each other's lives
means the good and the bad, and we certainly have done that for each
other. We've laughed and cried together, and tonight one of us is
hurting big time . . . . Linda's son committed suicide last night. Mary
Williams is helping Linda [not Linda's real name]. I don't know Linda's
number, so I haven't spoken with her. Virtual hugs and cinnamon rolls
seem so futile.

  To:  Bev Sykes
From:  Nancy Norton

Oh no!!!!!!!

Thank you so much for posting this. I never met Linda in person, or
spoke to her on the phone, just exchanged a few messages with her, but I
find myself crying as I type this. I want to send big hugs not only to
Linda, but to everyone else here. If you're all as hit by this as I am,
I know you'll need them.
```

Linda's son lived in a different state than she did, and she was distraught that she was unable to go there immediately. So it was her virtual friend, Mary, who lives where Linda's son did, who dropped everything and did all the errands and legwork locally to help Linda with her son's funeral arrangements. A virtual relationship, true, but as real as they come.

# Everybody's favorite online service: e-mail

The dilemmas of long-distance relationships aren't new, but e-mail provides a solution that's the next-best thing to being there. It makes the latest calling plan gimmicks sound old-fashioned and expensive no matter what the price, even though e-mail goes through phone lines too.

## Family e-mail trees

Remember phone trees? You know—the way each person called three other people, so pretty soon you'd notified everybody in the club about the meeting next week, in no time flat? Kathy Harradine, of Concord, California (shown in photo), says that's how what she calls her family e-mail tree works.

When she was a child, it was her family's custom for all the members to add their news onto a letter, and send it to the next relative in the chain. "So by the time you got it, it was this huge package," she recalls. Partly because of that tradition, about 100 members of her extended family still keep in touch regularly. But today, most of them (even some of the oldest) communicate by e-mail. "It originally started as a phone tree, because a lot of us are single parents. Our kids have always known whom to call if they were in trouble, and who would call whom next. So the family's always had a way to connect."

As their earth-mail tradition evolved into e-mail, Kathy and her relatives joined whichever online service offered a local-access phone number where they live. That's why various family members ended up on America Online, CompuServe, GEnie, Prodigy, and the Internet. But that makes no difference, because anyone can send e-mail from one service to anyone on any other service, often for no charge other than their monthly membership fees. And receiving e-mail is free on most services, just like receiving earth mail is free. For outgoing mail, it costs less than first-class or air-mail postage, even to another country. Better yet, it arrives almost instantly.

> *Within an hour-and-a-half, tops, the phone had stopped ringing, and everybody knew that Brian and Colleen were just fine.*

The Harradine clan had used their e-mail tree to send invitations to a 50th wedding anniversary celebration and holiday letters and ordinary how-are-you letters, but the real test came the night in October 1994 when a race riot flared then fizzled in Kentucky, after a white cop shot a young black man. Kathy's son, Brian, and her new daughter-in-law, Colleen, had just married and moved across the country, to Lexington. So when relatives heard the news reports of the riot, about 25 of them called Kathy, one after the other. Nobody could reach Brian and Colleen by phone to make sure they were safe, because all long-distance circuits into Lexington were busy, as is often the case in any crisis or natural disaster. Kathy short-circuited that problem by sending e-mail to a cousin in the Lexington area who could make a local call and report back right away. She then sent his "all's well" reply, by e-mail, to the relatives in the top tier of the family e-mail tree. Those people already knew their assignments, so each of them relayed the message to a few more relatives.

"Within an hour-and-a-half, tops, the phone had stopped ringing, and everybody knew that Brian and Colleen were just fine," Kathy says. Aside from that first flurry of calls to Kathy before she sent the e-mail message, no one paid more than whatever they would for a local call.

Another advantage of e-mail, says Kathy, is that it's a lot cheaper to pay less than ten bucks a month for the basic fee for an online service so your kids in college can send you e-mail than it is to pay for collect calls, or to call them. "Besides," she says, "did you ever try to actually reach a kid in college? This way, they pick up your message when they have time and reply when they can, but at least you do finally hear from them."

Connie Strand, of Kansas City, Missouri, says she now keeps in touch with a cousin regularly by e-mail, whereas she used to talk with her only about once a year, at the family reunion. "One of the things I like about e-mail is that you still have to write, and that people can still do that, despite all the TV they watch. E-mail is kind of between a phone call and a letter. I don't have the time to sit down and write a letter, but I can take the time to dash off two or three lines in an e-mail message."

Lisa Kimball, who runs The Meta Network in Arlington, Virginia, says the same idea applies to both friendships and business. "I can have friends in Japan with whom I can stay when I'm there, and have them stay with me when they're here. Or instead of just meeting someone once a year at a conference, I can exchange e-mail with them all year, and have a real relationship."

## Your very own mailing list and fan club

Even though most people meet others online in forums where they have common interests, they soon shift to e-mail, which is where the real business networking goes on and the friendships develop. Rosalind Resnick, of Hollywood, Florida, has become friends with many of the business associates she has online, so she came up with a creative way to turn the technology to her personal use as well.

*I'm much more turned on by swapping e-mail with people I know and people I like than by chatting with strangers in forums.*

Since way back in her grade-school days, Rosalind says she's always been the type who invented her own game while everybody else was playing the predictable ones. So she started her own private Internet mailing list called Friends of Rosalind. "I'm much more turned on by swapping e-mail with people I know and people I like than by chatting with strangers in forums," she says. "The only rule we have is that we never talk about work." People who receive the mailing list send her essays about themselves to bring others up to date, then Rosalind writes a newsy report to add to the mailing. The list of participants has grown so much that she automated the process by using listserve software, but she began simply with group e-mail.

You can do that too. It's especially easy with the automated software available for most services. Almost all software enables you to save people's online address with just the click of a key or two. You usually have the option to save them in groups, too, which creates an automatic group e-mail list. Just check the manual or the built-in Help file.

## Marriage, modem style

E-mail is also what makes being temporarily apart easier for Frank and Barbara Meissner. Frank's an Air Force doctor, and it was while he was based in San Antonio that Barbara began to fulfill her dream of getting her master's degree in anthropology. Then Frank was offered a promotion as head of the cardiology unit at an Air Force medical center in California.

There they were, both with good opportunities, but in different states. He went and she stayed. Until she finishes her graduate work and they can be together again, they keep in touch daily by e-mail. Both say that Frank works such long hours that even when they were under the same roof, they seldom had much time to spend together. "We've probably spent as much time actually talking to each other since he left—most of it by modem—than we did before," Barbara says.

E-mail has also revived the nearly lost art of letter writing, and resulted in so many couples courting by correspondence that Elizabeth Barrett and Robert Browning now look like trendsetters. Many of these techno-era romantics have eventually met and married, and lived as happily ever after as those who first met in traditional offline ways—maybe more so, because at least they took the time to get to know each other mind-to-mind and soul-to-soul first.

# Safe and easy ways to make friends online

The best way to make friends online is the same way your mother taught you to make friends when you were a teenager. Sandra Hibbs, a former ballet teacher turned Internet instructor for America Online, puts it this way: "Stay away from personal conversations to begin with, just like you do in the real world. Develop your friendships through mutual interest, rather than through some internal need for friendship or romance. It's better if you just start communicating [in forums], where people are there because they have a common interest."

Other tips: Never post your phone number or address in public, and avoid mentioning your neighborhood or exactly where you work. When you start exchanging e-mail with people, wait until you get to know them before you give them your phone number or address, and never give your credit card number to anyone in e-mail unless you're absolutely sure the person is legit. It's safer to use your credit card for purchases in designated online shopping areas, but even that carries a bit of a risk.

## Yeas and nays about chat rooms

I admit it. Personally, I think paying by the hour to type banal, three-line messages to someone, stranger or not, is a waste of time and money. I also dislike the online "chat invites," where someone interrupts whatever you're doing to type inane things at you such as "Where are you? Do you know how to do this?" To me, that's irritating and intrusive. It's like a stranger coming up to you on the street and trying to strike up a conversation. Unless it's an online conference featuring an expert I wouldn't have access to otherwise or perhaps a quick business consultation with a colleague who's halfway around the world, an online chat is cheaper than a phone call. If I want to carry on a real-time conversation, I want it to be with someone I already know or have a reason to talk with. Most times, I'd rather pick up the phone. But many people think online chats are terrific, so to each her own.

Contrary to popular perceptions, it's not just people with empty lives who can't normally communicate well in real life who hang out in chat rooms. Some are, sure, but bear in mind that the reason they can't communicate verbally might be because of a speech or hearing problem. In that case, online chatting might open up a whole new world, and is undeniably a great boon. Sometimes, too, those who engage in real-time chats are people like Amy Bernstein, who certainly scores high enough in communication skills because she's an associate editor at *U.S. News & World Report* and the founding leader of the Women's Forum on CompuServe. When she was first online, Amy says she spent a lot of time in the chat areas.

"When it was a novelty it blew me away. It was cocaine! I felt that it was such a powerful tool, because I could talk to perfect strangers in Rapid City, South Dakota, or Mexico City, Mexico. . . . When the novelty wore off, it stopped being fun. I got tired of slow-motion conversations and not being able to see or hear the nuance in what someone said, and it's not as spontaneous as a phone conversation. So now I use it strictly for work. But I met two or three people that way who are still very dear friends."

In her early years online, and occasionally since, Sandra Hibbs ignored her own advice about keeping to public places with a purpose, and spent many hours in chat mode, both one-on-one and in groups. "I like the growth aspects of it," she says. "It's like being in a huge room with everybody discussing things from a different point of view. It broadens your mind and gives you a different view than your own. One time we spent four hours online creating a utopia. Of course, everybody's idea of utopia was different."

CompuServe, eWorld, and some other services require real, full names in forums, which prevents most of the crude, rude, and lewd messages that people sometimes send when they can hide behind the mask of anonymity. But in chat areas, such as America Online's People Connection rooms or CompuServe's CB Simulator, people have the option of using handles, or fake names, and most do. There are ways to ward off offensive messages if you want to explore those areas, though. The best way is simply to ignore any invitations to chat in private rather than in the public chat area where others are also present. Even then, Sandra says, "If you're going to survive in the chat areas [without getting unwanted sexual come-on messages], it's a good idea to start talking about specific topics or special interests, like movies. If everyone starts to contribute or bring up a new idea, it really starts going. Don't just sit there, become a participant."

Sandra once spent a long time getting to know a person who used the name Tupperware, who was another regular in one of the chat rooms where she hung out often. Yes, Tupperware. "I thought it was a female for about a year," she says. "I finally met him and found he was a *he*, an engineer who worked on the space shuttle; he chose that name because one of the materials they used was like Tupperware."

# CHAPTER 5

# For parents, about kids

This story really began months ago, way before the New Year's weekend that I got involved. It has no ending, because it's about real life, and real-life stories don't divide easily into chapters or always come to some satisfying closure. A lot happened to a woman named Karen between the summer day when she first posted a call for help to the misc.kids newsgroup on the Internet and when she came back and posted another during the holidays, and as a result of what she revealed about her life, many other people's lives were affected too. Most of them—most of us, I should say—have never met face-to-face, yet had we not met in that hypothetical realm of cyberspace, our lives would be a little less rich. What was said has stayed in our minds and in our hearts, and it's a practically perfect example of how life can be online, at its best. Karen's first message, posted publicly in that newsgroup, read (with her name and identifying details changed):

"I have a rather personal request. My husband is asking for a divorce, and there seems little doubt that this will occur. His main complaint is that we have not been intimate much in the last three years. Without going into great detail, it is true I have not been terribly interested. My response has been that I have been either pregnant or nursing an infant for nearly 3½ years. (All but two months. We have a 2½-year-old, a one-year-old, and I am now eight months pregnant.)

"I feel inadequate and I question whether or not I am normal. My husband is a psychiatrist, and tells me I have a problem that will not go away when I stop nursing and having babies.

"I don't expect to change my husband's mind. After all that has been said, I don't want to. But I would like to understand myself better. Am I making excuses for myself? Or am I normal in not having much of an appetite while nursing or being pregnant? I want to hear from all of you (Moms and Dads), not just those who empathize with me."

Many people, both men and women, responded. The women talked of how tired they had been while they were pregnant and nursing, and the men told how little they had understood about what their wives were going through during similar times. Others offered different insights:

"Lots of responders have mentioned tiredness as their cause of lack of desire. I experienced the same thing, but I don't think I'd label it as a result of exhaustion. I just didn't want intimacy. Maybe it was because I was giving so much of myself, what little I had left I wanted for me," said one.

Another woman reported: "My husband and I have gone through our share of this as well. We have three kids. The youngest just started sleeping through the night, and the middle one has been waking up nightly for about six months now. It takes a lot of energy to maintain a house and family, let alone get in that quality time. We both work full-time as well. Getting the sleep is often preferable to me than sex. It hasn't always been this way, and I know it won't stay this way. And it in no way has anything to do with my husband."

Some suggested scheduling date nights and others recommended counseling so they could try to save their marriage for the sake of the kids. Some took Karen's side; some took her husband's side:

"If your marriage is important to you, sex is something you have to commit to making time for. I didn't understand until recently how seriously important the need for sex is for many men to feel that they are loved. . . ."

"Well, the one-year-old and the baby on the way didn't get there by magic. Quite frankly, I can't imagine a woman in your situation right now having much sexual desire, mainly because you are tired."

One woman suggested Karen go out of town on business and leave her husband to juggle his job, two kids, and the whole household for a week and see how sexy he felt after that, even without being pregnant. She also commented: "The thing is, in addition to the tiredness (which is bad enough) it sounds like there is also some (well-deserved) resentment on your part, which could really get in the way of intimacy of any sort."

That discussion went on in mid-summer. During the week between Christmas and New Year's, Karen once again left a message in misc.kids, but this time it was more a cry of pure anguish than a plea for help because she knew there was nothing anybody could do to help her change what had happened. The baby that she was still pregnant with when she posted her first message had been born in the fall, and had just died of SIDS (Sudden Infant Death Syndrome).

"I have never known such profound grief," she wrote to the group this time. "My arms are achingly empty. The two weeks since she died have been the hardest weeks of my life. . . . I have no idea why I'm posting this. Perhaps I'm looking for someone who has been through it." Later, after a few replies, she added: "A part of me so wants people to understand the sheer depth of this pain that I'm willing to grieve forever to prove it. This ache, this longing, goes beyond a 'recent sadness.' It cannot be 'wiped away.' When talking to a grieving parent, don't allow yourself to trivialize. The sweetest people on earth do it all the time. Everything is not okay here, and it isn't going to be for a very long time. That has to be acknowledged. . . . The very kindest thing people can do for me is to be willing to listen to me, willing to see my tears, willing to bring it up through the use of her name. . . . Cry with me. That't all you can do."

Now I don't even have children, not anymore at least, so I don't know why I was browsing through messages in a newsgroup about kids around midnight on a holiday weekend. But I happened upon a reply to Karen's note, and read through the other replies to her. It was the ones with the theme of "your baby's probably in a better place" that triggered my reaction. My only child had also died of SIDS years before, and those responses made me remember how much those kind of comments had hurt and how angry they made me at the time. So I left a message for Karen. I don't even know what I said, but toward the end of writing it tears were streaming down my cheeks and it felt good to hit SEND.

It felt good because I'd been able to help another mother whose particular pain and grief I knew well, so it helped me resolve a bit of what lingered of my own. Over the next couple of weeks, several people, both dads and moms, sent me private e-mail messages saying how much the exchanges had affected them too. It was one of those women who told me about Karen's message months before, about her husband wanting a divorce because she had so little sexual desire, which made her situation even more poignant to all of us. Another wrote me to say that she had lost her first baby to SIDS and now had a three-month-old baby daughter. She worried about whether Karen would blame the children for her marital problems or blame her husband for the baby's death, and feel guilty for years either way. And because of what happened to Karen, her own fears about losing her new baby to SIDS, as she had her first one, had resurfaced. From my own experience, I knew that isn't the kind of worry that SIDS parents say aloud to each other, much less to other relatives or friends, because you're almost afraid that saying it might make it true. It becomes an unspeakable fear, yet it's there, eating away at you. So I like to think that expressing her thoughts to a stranger in e-mail helped this woman too, and I tried my best to reassure her that everything would be all right. Anybody who has ever lost a child in any way never trusts such platitudes again, and we both understood that. Still, it was like getting and giving a hug.

The point of this story isn't just its poignancy. It's to show you that, contrary to the beliefs of many people who haven't been online yet, it's not all tech talk and certainly not "sterile and mechanical" as the woman I quoted in the introduction assumed it would be. Nor is the nature of this story unusual. It's merely a snapshot of what goes on all the time, because the people online are real and they talk about real things.

# Get resourceful, get results

Whether it's a routine need for advice or information or a real crisis, many people have found resourceful ways to use online services "in real life," or IRL as some people say online. But as wondrous as the resources can be, you might need to figure out creative ways to use them to do what you need to do, as Lee Graham, of Tucson, Arizona, did.

Shuffling through the morning mail one day, Lee found a letter addressed to her 16-year-old daughter that forced her to confront the fear that all parents hope they'll never have to face. The letter looked like it came from a prison, and was from someone she'd never heard her daughter mention. Lee panicked. She felt guilty about invading her daughter's privacy, but she opened the letter anyway.

That letter turned out to be from a man serving a life sentence for killing three children, but Lee didn't learn that part until later. A few days after she questioned her daughter about the letter and her involvement with this sinister stranger, her daughter and a girlfriend of hers ran away together. It was when she started digging frantically for information that Lee discovered that both of the girls, along with several other teenagers and adults, were involved in a Satanic cult led by that convicted murderer. Lee, however, reacted with the logic and resourcefulness of a good detective.

"I searched her room, and found lots of phone numbers on pieces of paper. If it's a listed number, Phone*File on CompuServe (GO Phonefile) gives you that person's name and address. That turned out to be real useful.

"I called the people, and drove by and talked to some of them. If you call the police and try to get them to track a number, you won't get much help, especially if it's an older child who has run away. Phone*File charges a very minimal fee. I probably spent all of $7 to establish the pattern of what part of town to look in to find her, then to figure out where they'd probably gone when they heard I was looking for them, and moved on.

"Then I used the Newspaper Archives (GO newsarchives) to look up information on the case of the man who sent the letter. That's how I knew I was dealing with a truly serious situation."

From her several years online, Lee knew that there was a section or forum for just about anything, so she logged on and typed FIND missing. That led her to the Missing Children's Forum, where she immediately found other parents who had dealt with similar crises. They gave her much-needed emotional support, as well as practical help such as the phone number for the Cult Awareness Network in Chicago.

Day after day, for five weeks, Lee continued her search for her daughter and her friend. "I took flyers with photos to the Circle Ks and the convenience stores, and asked them to watch for the girls. Finally, the Circle K people reported that they'd seen them, and they were picked up in Catalina, the town where I thought they would be. But I would never have found my daughter without CompuServe."

When I last talked with Lee, with a deep sigh she said, "I'm not sure that I won't lose her again. She's got a lot of problems." But she had just learned about Medsig, with its Mental Health Section where psychiatrists and psychologists hang out daily, so that was to be her next stop.

# The risks of letting children and teenagers online

In Lee Graham's case, an online service played a major part in the solution to getting her child out of danger, but they can occasionally be the problem. Yes, it's true that pedophiles lurk online and try to engage kids in sex-related discussions by e-mail or in real-time chats, or sometimes try to lure them into meeting them somewhere offline. On the Internet, there are even two newsgroups where they're likely to be lurking (alt.sex.intergen and alt.sex.pedophile). According to the National Center for Missing and Exploited Children, teenagers are more vulnerable than younger kids, because they're more likely to hang out in the places where people discuss relationships.

However, the risky situations are rare, especially on the commercial services. On most major networks, such as America Online, CompuServe, GEnie, and Prodigy, there are ways to block access to certain sections or mail from specific ID numbers. Yet here too, as Lee Graham had to do, parents have to choose between allowing their children some privacy and autonomy, and watching their every move.

## How to block access to services or features

*America Online* Click on Member on the main-screen menu bar, choose Parental Control, and choose Block Instant Messages. This can be done only with a master account, or the screen name you used when you joined the service.

*CompuServe* Send a message to *Sysop in any forum that you don't want your child to enter, and ask that your ID number be blocked. That means you won't be able to enter that forum either, unless your children have separate accounts. To block access to CB Simulator, the giant chat room area, send the message to *Sysop in the CB Simulator Forum. CompuServe will also introduce special software soon, called KidNet, so you can selectively block access.

**Delphi** Delphi has, blocked all access to a few Internet areas because of pornographic material.

**Echo and the WELL** In any conference, you can set any topic to Ignore at the Respond prompt, which makes it invisible thereafter, unless you type remember and that topic number. New topics are started all the time, however, so this would require you to constantly monitor all conferences. You can't block an entire conference, although you don't get to any conference unless you actively choose to go there.

**eWorld** Send a one-on-one message to $IM OFF, with anything in the body of the message. This blocks all one-on-one invitations. To reactivate, do the same, but substitute ON for OFF.

**GEnie** You can ask a roundtable sysop to block your ID number in that area, but you won't be able to get into that section either if you do.

**Interchange** Ask for instructions in Member Support.

**Prodigy** Jump to Household Access —> Add a Member —> Manage IDs —> Change BB Access. You can give up to six people in the household separate ID numbers and passwords, although they're all keyed to the A, or master, account. You can also restrict access to Plus areas, meaning those that cost extra, although that's true of virtually all the forums. Or you can choose just which forums to block access to, and make that different for each child. You can also block access to all chat areas.

**Women's Wire** You can delete the icon for any conference, but you won't be able to access it again if you do.

Always remember this: When you're online, you're in a public place, even though it doesn't feel like it because you're also in the privacy of your home. Just as they do offline, parents need to teach their children how to protect themselves when they're away from home.

Parents also need to resist the temptation to let the computer and modem become the babysitter, says John Lucich, an investigator for the Organized Crime and Racketeering Bureau of the State of New Jersey Attorney General's Office, who also runs the High-Tech Crime Network BBS. "While I believe that sysops have a responsibility for what goes on on their networks, the ultimate responsibility for what goes on with children is the parents," he says. "We put alarm systems on our homes to lock criminals outside, yet we put modems on computers and let them play with criminals online.

"We can't let our kids lock themselves in a room for hours and let them play on the computer. Put the computer in an open area of the house. Even then, they could go to a friend's house where they can use the computer out of the parents' sight. So the key is more open communication between kids and adults."

– 83 –

## My rules for online safety

(Copy, and keep this pledge at your computer.)

- I will not give out personal information such as my address, telephone number, parents' work address/telephone number, or the name and location of my school without my parents' permission.

- I will tell my parents right away if I come across any information that makes me feel uncomfortable.

- I will never agree to get together with someone I "meet" online without first checking with my parents. If my parents agree to the meeting, I will be sure that it is in a public place and bring my mother or father along.

- I will never send a person my picture or anything else without first checking with my parents.

- I will not respond to any messages that are mean or in any way make me feel uncomfortable. It is not my fault if I get a message like that. If I do, I will tell my parents right away so that they can contact the online service.

- I will talk with my parents so we can set up rules for going online. We will decide on the time of day I can be online, the length of time I can be online, and appropriate areas for me to visit. I will not access other areas or break these rules without their permission.

By Lawrence J. Magid. From the "Child Safety on the Information Highway" brochure, copyright 1994, National Center for Missing and Exploited Children (NCMEC). Reprinted by permission. Full text available on most online services. On CompuServe, go to the Missing Children Forum Library, filename CHISAF.TXT. This brochure was sponsored by the major online networks. You can also get a free copy by calling 1-800-FIND-LOST, in the U.S., Canada, and Mexico.

# The rewards for kids online outweigh the risks

When both parents and kids use online services responsibly, the rewards far outweigh the risks. Prodigy has long been the most kid-oriented, with excellent, fascinating activities, message boards, and libraries. But America Online is catching up, and CompuServe, eWorld, and GEnie have attractive features too.

Imagine your children's delight when they discover "Sesame Street" or "Nova" online (see Fig. 5-1) or check into *National Geographic*'s online activities on Prodigy to find that this week's feature is a trip to Mars, complete with graphics and sound. "It is the year 2015," reads the opening screen. "You are the commander of the first mission to colonize people on Mars." Your spacecraft take-off has been delayed by three days due to rain, it goes on to say. Today, it's sunny, but it's also Friday the 13th and there's more rain predicted for tomorrow. So is it Go or No Go? Your child gets to decide, then contend with the consequences of that decision.

**5-1** *Nova's Mother Nature, Inc. interactive science education service in Just Kids on Prodigy.*

There are also a multitude of things that are endlessly fascinating, no matter what your age, on the Web. To get the full benefit of the sites with animation and sound, you'll need a fairly fast and powerful computer with a sound card, external speakers, and a 28.8-bps or faster modem. Even without all that, as long as you have Web browser software (see appendix B), you can still see the graphics and find everything from a frog dissection in process to photos of paintings in the Louvre. The Web can take you almost anywhere that's on the Internet, worldwide.

## The father and daughter behind Kids' Web

If you read the Introduction, you might recall that the norm has been for dads to buy computers for their sons, and for parents to encourage sons to use them more than they do daughters. Not so with Steven Burr and his daughter, Emma. Steve created a Web site called Kids' Web with Emma's help, and linked it with the other Web sites they had explored together that were Emma's favorites. It also includes Emma's favorite works of art, links to educational sites, children's poetry, children's software recommendations, and just plain fun sites on the

Web. The result, intended for children ages four to ten and perhaps beyond, is a great starting point to introduce young children to what's on the Web and teach them how to navigate. All they have to do to get from Kids' Web to, say, the Paleontology Museum at the University of California at Berkeley or the White House is point to the appropriate reference on the Kids' Web screen, click the mouse, and zap! They're instantly transported.

"My main purpose in exploring the Internet with Emma is to begin to get her comfortable with the concept of obtaining and sharing information by communicating with remote computers and computer networks," says Steve. "I believe that people who do not have this skill will in the future be at as great a disadvantage as people today who are not good readers.

"The second but nearly as important purpose for spending time online with my daughter is just to spend the time. Emma and I share an affinity for the computer that others in the family lack. My wife doesn't care for it much, and so far my three-year-old son refuses to touch the mouse. So it's a very special time for us, and we have a great time together. We concentrate on the World Wide Web, because it is the easiest and most entertaining of the Internet tools. But ultimately I want her to learn how to use all of the Internet resources. . . . I hope Kids' Web encourages other parents to do similar types of things with their kids."

Emma, who was all of five years old when Kids' Web went online, says, "I like the Internet because it's fun, and I especially like the White House. I also enjoyed learning about the dinosaurs." Her more candid view of the value of Kids' Web is just as practical as her dad's, but in a different way. When Steve told her that there are as many as 30 million people on the Web, and a thousand people visited their Kids' Web creation in just the first month, Emma's response was: "It's good that a lot of people will know about me. I've always wanted to be famous."

The Web has countless sites that are both educational and entertaining to visit, and can be a great way to do research for school projects, besides stimulating and satisfying a child's infinite curiosity. The more interactive it gets, beyond the passive process of reading or just pointing and clicking, the more appealing it will be to younger children. There's already a range of sites, from sound and video clips and text pages created by bands to graphical information from NASA, that will entrance all but the most jaded kid for hours. If you can find a way to share the experience, as Steve and Emma have, so much the better.

Most commercial online networks have special areas for younger kids with games, clubs, special events, projects to do offline, and stories that involve problems for them to solve or decisions to make. Several also have advice features run by adults, where kids can ask questions that they might be embarrassed to ask you, as well as places where they can talk things over with their peers.

These areas are supervised by staff and volunteers screened by the service or the forum leaders, so they're safer than letting children loose on the Net, especially if you've used the features each of the services have for parents to block access to certain sections (see the shaded area earlier in this chapter). If you ever feel uncomfortable about a message, you can easily delete it and block any future messages from that individual by using his or her ID number or name. Report anything that really concerns you to the network's staff immediately. That said, there's no reason for undue alarm. All commercial services are acutely aware of the problem and are vigilant, so it's highly unlikely that your child will encounter any problems online.

You'll find more information about what's online that's especially interesting and useful to students in chapter 9, *Lifelong learning*, as well as resources that older students can master in chapter 18, *Online research basics*.

# Highlights: Places for kids

## For children

America Online has a Kids Only section, as well as a Time for Kids section in *Time* magazine's forum; CompuServe's S-Net, or Student Forum A, has sections for elementary school-age kids. GEnie's Family and Personal Growth Roundtable has a section called For Kids Only (through age 12), which is accessible only upon request and approval of the sysop, and they also have a section called GEnie for Kids. Prodigy, as mentioned previously, has an excellent Just Kids section within the Family/Home area, plus Sports Illustrated for Kids. And there's KidsNet in the TV and Radio Forum on AOL.

## For preteens and teenagers

Nobody's neglecting the interests of older children and teens, either. The major commercial nets have forums for students, and of course the forums about movies, music, and a myriad of other topics will most likely appeal to teenagers. Other than the forums created mainly for help with homework or an opportunity to communicate with peers, the best of the lot are the interactive forums on Prodigy sponsored by Nova, NASA, and the publisher of the *Babybsitter's Club* series of books. Advertisers often hype something as "interactive" when the only interactive aspect of it is that you have to turn the machine on and maybe point and click. The services that are truly interactive actually involve the kids in making decisions and give them feedback based on the answers they give. They teach everything from science facts to ethics and responsibility in the process, and often provide a forum to discuss the decisions with other participants.

Another exemplary example of how this medium can be used well are the opportunities both through the Internet and through some of the commercial nets—again, mainly on Prodigy—for children to communicate with astronauts, anthropologists, and other explorers while they're actually on expeditions by leaving e-mail that's relayed to them. The explorers also send back periodic reports on what they're finding, and the sections include plenty of background information on the expeditions. It's an opportunity for indirect but active involvement in history-making events, rather than just passively reading about them or watching news reports on TV.

## For college students and young adults

CompuServe's S-Net, or Student Forum B, is for the college-age crowd; Delphi's Custom Forum 312 is for Generation X; BITCH on Echo and FemX on the WELL are for younger women.

# Highlights: Places for parents

## On the commercial networks

Each of the commercial services has forums or sections about parenting, either separately or as part of home, family, or lifestyle forums. The keywords *parents* or *parenting* should lead you to them. Some examples: the Parents Information Network on America Online; The Missing Children Forum (*GO missing*) on CompuServe, sponsored by the National Center for Missing and Exploited Children; and on Delphi: Child Care Support Net, Custom Forum 196; The Family Forum, Custom Forum 092; Kids First, Custom Forum 208; and Single Parents Network, Custom Forum 007. On GEnie check the Family and Personal Growth Roundtable, and on Women's Wire look for the Parents' Clubhouse section in the Parenting & Education section of Exchanges (see Fig. 5-2).

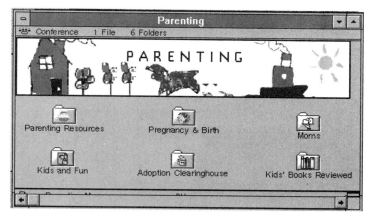

**5-2** *The Parenting section of Women's Wire.*

## On the World Wide Web and the Internet

The National Center for Missing & Exploited Children maintains a Web site at http://www.scubed.com:8001/. For information on adoption of non-U.S. children, including photos and brief biographies, check the Web at http://www

.gems.com/adoption/. There's also an Adoption Network for adoption agencies, support groups, publications, and attorneys, which could be a good source for further information. It's on the Web at http://www.adoption.org/adopt. Among newsgroups, check misc.kids, one of the better newsgroups on the Net, and alt.parents-teens.

# CHAPTER 6

# Sex and romance

**S**urprise, surprise. According to a *Wall Street Journal* article, *sex* is one of the most frequently used search terms on the Internet. Much of the buzz in the media is about cybersex, also known as *compusex*, but by any name it's the online equivalent of phone sex. Some call it *safe hex*, a pun on the computer term hexadecimal. To whatever extent this is true, it's only one factor in why people indulge in typing erotic messages to one another.

# The appeal and risk of cybersex

Gloria Brame, coauthor of *Different Loving* (Random House), says that most people who engage in cybersex are married or in committed relationships and not likely to cheat on their partners in real life. They might be motivated to experiment online because of something as simple as a partner who refuses to consider oral sex, she says. "It's not that cybersex is exciting because it offers opportunities to people with unusual fantasies. But they might have fantasies that they'd be embarrassed about experimenting with in reality or no opportunity to explore in reality. The anonymity is extremely liberating. They can talk about things they would never share with people.

"A lot of it has to do with the power of the imagination. It's like custom-tailored porno. Just seeing the words on the screen is very powerful, because they know that those words are directed to them. Sometimes I think it's part of the online theater."

Howard Lewis, who coleads the Human Relationships and Sexuality Forum on CompuServe with his wife, Martha, echos what Gloria says about the nature of cybersex. "It's interactive pornography," he says, "and emotionally engaging in unexpected ways." Therein lies part of the problem. Both Gloria and Howard say that the real risk is when it leads to phone conversations and then offline rendezvous, as it often does.

"It's not the same as renting a porno video, because there's a human being on the other end of those pixels," says Howard. "That can be threatening to a marriage. If it starts leading to things outside the understanding of the marriage or if it causes any difficulty with their partners, they'd better discuss it, and may need to forgo the thrill."

The thrill wears off pretty fast anyway. "Sooner or later, if it doesn't turn into something else, people get bored with it just the way they do in real life. A lot of people do find it plain silly, but a significant percentage of people get very caught up in it," says Gloria.

"We're told that after the initial thrill, many people find it rather tedious," says Howard. "Everything sexual is potentially addictive and potentially boring."

# Revelations about online romance

Not all online romances involve cybersex, however. Far from it. Yet it's still usually difficult, if you haven't experienced it, to understand how you could get so enamored of someone you meet online. And it's rather amazing how quickly you can become obsessed with someone whom you know only from words on a screen. It shouldn't be, really, because there's nothing new about courtships by mail, although it was nearly a lost art until e-mail became so widely available. However, it's not just the method of delivery that's different. E-mail is a cross between written and verbal communication, in that it combines the time to think about what you want to say and how you want to say it as writing does, but it feels more like the instant communication of speech. You don't have to wait days or weeks for a reply, either. You could get an answer within the hour, or almost surely by tomorrow if it's a hot-and-heavy correspondence.

Another thing that sets communicating by computer apart from either purely verbal or face-to-face communication is that your entire attention is focused on the words and the thoughts behind them. There's no voice to distract you, but no tone of voice to give you the kinds of cues we rely on to hear how people mean what they say. There's no monologue running in the back of your mind about how attractive the other person is or isn't, or you are or aren't. Nor is there the sideways glance, sheepish grin, nervous twitch, nor flashing eyes or lowered eyelids to reveal telltale signs. Online you lose all that, but gain something too, because it's purely mind to mind and soul to soul. And who doesn't yearn for someone who appreciates them just for who they are, inside? How seductive!

Thus it's no wonder that many perfectly rational people get seduced online. It feels so anonymous that you're not as guarded as if you would be encountering that person face to face. So it feels perfectly natural to confide in someone sooner than people normally do, which fosters instant intimacy.

Most people start these relationships casually, intending no more than brief banter, and most of the time that's all that happens. Yet the intensity and the intimacy can change all that faster than anywhere else other than in the movies, or perhaps aboard a ship or during a week at Club Med.

Online infatuations are so common that at any given time in any online area open only to women, you can probably find one woman discussing whether to actually meet some man she's been corresponding with and fantasizing about, and the other women trying to talk her out of going further and warning her that online romances rarely work in reality.

Kathy summed up the experience of many who've been there as she tried to talk some sense into Joan, who was considering a real-life, clandestine tryst with her newfound online lover—what people in the Human Sexuality and Relationships forum on CompuServe call a *3-D meeting* and others call a *face to face*. It had to be on the QT, you see, because the guy was already engaged to someone else. But that's not unusual, because many of the men who strike up online romances turn out to be married. Another big surprise.

"Online affairs *always* take off fast," Kathy told Joan, "Time is condensed online. We meet mind to mind, have great conversations with someone about subjects we'd never discuss offline, and feel we've met our soulmates. It's not an unusual occurrence here. So it's not specific to your relationship with this man. I mean, you met this guy a week ago, exchanged sexy e-mail and phone calls, fully expect to be in bed with him when you meet, and seem to be fantasizing a long-term relationship with him. Do you feel any offline affair would progress so fast? Whether he should or should not be marrying when he's making plans to cheat on his fiancée is his problem, not yours, but are you sure he's not also involved with others?"

Kathy's point is quite valid, says Howard Lewis. "A lot of these men are just Lotharios. Certain women are vulnerable through the fantasy element of this medium. We urge women to be extremely cautious of online relationships, because you really don't know very much about somebody, and there are some who are predatory. Women also have to be wary about being hounded by people who are mooning over them. Their lives are not threatened, but they feel kind of spooked or uneasy. There's a feeling that somebody is intruding on their lives."

## How to avoid sexual harassment online

The best way to avoid sexual harassment online is to avoid hanging out in chat areas (which some people treat as the online equivalents of pick-up bars) or in self-help areas where the nature of the disclosures and discussions invites intimacy. The best thing to do if you do happen to get a lewd or lascivious chat or e-mail message is nothing. Don't respond at all. You can also hit Delete to wipe out the message, or use the squelch or bozo feature available on most commercial services and local BBSs to block any future messages from that person or ID number (see *How to block instant chat invitations* in chapter 1). Just like obscene phone callers, what these people want is a reaction. If they don't get one, they'll soon find other prey.

It's nearly always easy to fend off unwanted advances or anything even close to harassment. There are also several precautions you can take to prevent problems:

- Keep the discussion on neutral, business-, or hobby-oriented topics.
- Don't post your phone number, address, or workplace location publicly; give it only privately, in e-mail, once you've gotten to know a person through public forums first and for a while in e-mail after that.
- Stay out of one-on-one chat sessions and avoid chat rooms unless it's a business- or hobby-oriented forum. Even then, keep the conversation

focused on the topic of the forum, and be as prudent as you would offline in a public place.

- Always bear in mind that you're in a public place when you're online.

- Most people don't intend their messages to be threatening or truly intrusive, even if they're flirtatious. So you can usually convey "back off" without being that blunt. But if Romeo or King Leer doesn't get the message the first time or two, be blunt and be clear. "I'm not interested, and please don't send me any more messages" usually works.

- Remember, you can turn off your computer whenever you choose.

- If all else fails, follow Prodigy's rule: "Turn 'em in, and turn 'em off."

Gender benders, or men and women who masquerade as members of the opposite sex, either for kicks or out of curiosity, are another common problem in the forums that focus on personal or intimate relationships, says Howard. "We've had some very destructive experiences with female impersonators, because it really is a distortion of relationships. . . . There's such a feeling of mind-to-mind communication that when you find out the person isn't who you thought they were, you feel you've lost someone you care about, which can lead to enormous feelings of grief, as well as outrage."

> *. . . the experience of meeting someone offline whom you've known only online is like the long-used jargon about computers: WYSIWYG, or "what you see is what you get."*

More often, however, the experience of meeting someone offline whom you've known only online is like the long-used jargon about computers: WYSIWYG, or "what you see is what you get." Especially if the two people have been genuinely open and honest with each other, when they finally meet they find the other person is pretty much what they expected.

But a meeting of the minds alone rarely sparks real romance. There's that ineffable thing we call chemistry that ignites that spark. Most of the time when the would-be lovers actually meet, it's just not there. Yet in the course of researching this book, I heard from so many women who met their husbands online and are living as happily ever after as their neighbors who met in more conventional ways—which is to say probably about half the time—that I began to wonder if anyone meets in any other way anymore. Who knows how much AIDS, or the high stress and long hours of so many jobs, or the high divorce rates have caused it, but for better or worse one of the most popular ways to look for a mate in the '90s is online.

Lest all this sounds pathetic and ridiculous to the uninitiated, please bear in mind that most relationships that begin online are just like any that begin anywhere else. In fact, because people usually meet in a discussion area that drew them because of common interests, it's a heckuva lot better idea than singles bars. You also have a better chance of finding someone with whom you can at least sustain an interesting conversation than somewhere like a Parents Without Partners gathering, where the only thing you have in common is that you both have kids. But the place to meet people, either as potential friends or lovers, is in the forums, newsgroups, and regularly scheduled real-time conferences that have an agenda. Venture into the chat areas and you take your chances.

It's not just those who meet online who court online either. E-mail has been a great boon to long-distance romances, and has led to many marriages between couples who met in traditional ways but might never have had the time to get to know one another enough to fall in love or might have drifted apart because they weren't in the same city.

"It's a less threatening way to get to know each other," says Nancy Tice, a psychiatrist in New York City who spent much of her courtship and engagement period exchanging e-mail with Hal, now her husband, while he was studying in Boston to be a radiologist. "E-mail allows a closeness that's really pretty special. It's like the way it's easier to talk about things when you're driving in a car. It's romantic." And here's a twist that defies stereotypes: In her spare time Nancy is a sysop, and she was the one who taught Hal how to use his university Internet account to send e-mail.

# You can't get pregnant online, but...

Of course some people who look for sex-related information online are like Leah Ingram of Ann Arbor, Michigan, who simply wanted information on how to get pregnant. She knew how to get pregnant, of course. But when that hadn't happened a few months after she and her husband had decided they were ready to have a baby, Leah first sought her doctor's advice, as most of us would.

"But there were all of these questions that I had later that I didn't even know how to ask my gynecologist," she says. "The books I got didn't answer every question, and doctors don't have enough time. Besides, the questions I had were things that I felt silly asking anyone, like how does your cervical mucus change if you're pregnant, or do your breasts get sore immediately after conception or is that more likely just PMS, or why does my temperature chart look like a roller-coaster?"

By exploring the Internet through America Online, she found a newsgroup called misc.kids.pregnancy. "I asked the temperature question online, and 20 women wrote me to ask if I was taking my temperature at the same time every morning. The doctor had said to take it before I got out of bed, but didn't mention that it needed to be at the same time. So no wonder!" It apparently worked, because Leah's first baby is due about the time this book comes out.

"People share so much stuff with complete strangers," says Leah. That they certainly do, as this excerpt from a public discussion in the sexuality section of a women's forum shows:

```
I was wondering if I am the only woman in the world who rarely orgasms.
Also, anyone who can offer any advice please feel free."

There are 4 Replies.

No, you're not alone. And all I can offer for advice is DON'T FAKE IT!!!
I faked it for so many years that I have to fight the theatrical moan
now in order to keep my hubby going. But I'm learning and he's learning.
Mainly we're learning that there's no magical spot, and that it's a lot
of work. But what fun work it is! Another thing I learned is that you
usually have to figure out how to give yourself one, and then teach your
mate. And once again, practice makes perfect. And don't give up. It
takes awhile, but it's worth it.
```

# The 30 most common online questions and comments about sex

The following list is from The Human Sexuality & Relationships Forum Hotline on CompuServe (GO hsxtop):

 1  Is it normal to masturbate?
 2  How do I tell my child about sex?
 3  Female orgasm: what's real?
 4  What could cause impotence?
 5  What are the most common sex fantasies?
 6  How do you give a man oral sex?
 7  Can I increase my penis size?
 8  Why is my sex drive so low?
 9  Sex play with friend: "Am I gay?"
10  Is fetishism a sickness?
11  Does first intercourse have to hurt?
12  Will I run out of sperm?
13  Is there help for premature ejaculation?
14  Why does she feel pain with sex?
15  Will I always be a virgin?
16  How do you overcome shyness?
17  What are the stages of adolescent development?
18  How does aging affect sex in men?
19  Could a lump in my testicle be cancer?
20  How can I tell if I'm gay?
21  Why do testicles ache after petting?
22  How can you avoid yeast infections?
23  I dress in women's clothing.
24  How do you use condoms?
25  Safer sex: how can you have it?
26  What is the most effective birth control?
27  How often should we have sex?
28  I think I'm bisexual.
29  How do you help a victim of sex abuse?
30  What constitutes sexual harassment?

*People share so much stuff with complete strangers.*

As you can see, much of the talk about sex online ranges from frank to raw, depending on whether the forum or listserv is staffed by a sysop (systems operator, or online staff) or moderated, and how far you venture into the netherworlds of newsgroups or mailing lists on particular fetishes. That, of course, is up to you.

"It's a way for people to reveal who they are at a minimal risk of rejection," says Howard. "They can talk about things they can't talk about anyplace else in their lives. One time we had a real-time conference about methods of birth control, which was very quickly taken over by a voice out of the blue. A woman came on and said, 'I had this child, even though my husband didn't want it. I thought I would fall in love with it, but I don't like having a child, and I can't say this to anybody. You're not supposed to feel this way!' There was nobody else she could say this to. Other women expressed the ambivalence they had when their first child was born, or the problems babies can bring to relationships. This is almost unique to online communication, because you can get support from a very diverse bunch of people in different parts of the world, and people can say things that they can't say face to face."

# How one resourceful woman negotiated her divorce by e-mail

It stands to reason that if you can fall in love online and get information about getting pregnant online, there must be a way to use online networks to get divorced. Sure enough, Danelle Morton and her former husband negotiated most of their divorce agreement and worked through some of the emotional upheaval by e-mail.

"Everyone around me urged that I hire a nasty lawyer to savage Ryan [not his real name]," Danelle wrote in her own recounting of the story for the *San Jose Mercury News*. "Unsolicited, friends handed me phone numbers of lawyers whom they described in animal terms: shark, barracuda, junkyard dog, pit bull."

Rather than ravage her husband, Danelle wanted to resolve everything with as much fairness and compassion as possible, yet financially protect herself and their children. She found that phone conversations between her and Ryan quickly degenerated into tears, recriminations, and accusations. To complicate matters, Danelle felt that Ryan had a distinct advantage over her. "He's a professional negotiater, and those skills . . . made me feel like he was calculating," she told Robert Siegel in an interview for NPR's "All Things Considered." "In work and in my personal life, I'm not a negotiator. I tend to bail our or pout," she wrote later in her *Mercury News* story.

But her saving grace was that Danelle is a professional writer. Thus, instead of reacting to the emotion of the moment, communicating by e-mail allowed her time to think through what she said and how she said it, which enabled her to draw on her stronger ability to communicate in writing and counter his better strategic skills.

As she put it: "In contrast to what most people say—that e-mail encourages hot-blooded, sarcastic one-liners—for me it provided a style of communication that helped muffle the screams of outrage and imposed a structure of logic and rationality. . . . In my experience, when the situation is already very emotional, e-mail slows things down, diffuses the hostility."

So their e-mail dialogues helped them work through the toughest phases of coming to terms with both the legalities and the emotions, with a minimum of intervention and help from a lawyer. "The way we negotiated restored in me a sense of self-reliance I'd lost in marriage. But I wouldn't recommend it for everyone. For all my griping about how the marriage ended, Ryan turned out to be very reasonable in negotiation. And we had no horrible abuse or child-custody issues to resolve, which would have made this method unworkable.

"Recently he sent me a message that ended in XXOO. How strange, I thought, but I knew we'd made progress."

# Highlights: Where you can discuss sexuality online

On the commercial networks, look for the women's forums (see chapter 20), health and medical forums, and self-help or personal-growth forums. Most commercial networks also have sections within forums or full forums for gays and lesbians. Using sex or sexuality as keywords will lead you to most resources, but you might have to scout around a bit.

On the Internet, you'll find sex-related discussions in some of the misc. and soc. newsgroups about kids, parenting, and relationships, as well as the Couples-L mailing lists (send e-mail to listproc@cornell.edu, with the message SUBSCRIBE Couples-L plus your first and last names). Also, see the *Highlights* sections at the end of chapters 5 and 11 for suggestions.

For the truly adventurous, try any of the newsgroups in the alt.sex hierarchy (this is not one you want young children to check out). There are also newsgroups and numerous mailing lists for gays, lesbians, and bisexuals. You can find out about most of those either by sending e-mail to listserv@queernet.org with the message SUB gaynet plus your first and last names, or by visiting The Queer Resources Directory on the Web at the following address: http://vector.casti.com/QRD/.html/QRD-home-page.html.

# CHAPTER 7

# Domestic domain

**M**artyn Davis had been looking forward to this night all week. As a computer consultant in London, England, he hadn't exactly built his reputation on his culinary skills, but he was determined to impress his dinner-party guests and act like whipping out meals fit for gourmands was routine for him. He had invited his relatively new girlfriend, as well as his business partner who was bringing a woman he'd never been out with before, so both Martyn and his partner wanted to impress their lady friends.

"The bottles were lined up and the flat was cleaner than usual," he says. "I had bought a new set of dinner plates, which actually all matched. I had also done something I had never done before: At lunchtime, I went to a real butcher and bought a huge chunk of steak because the meal was to be a favourite of mine, Beef Wellington. . . .

"I got home early, put the meat on the kitchen surface, and reached for the cookery book where I remembered having seen the recipe. Then it hit me. You know how it is—time stood still while my vision tunneled down on the space where the recipe book used to be, before my ex-wife went away with it, the cats, and the marital strife. I panicked.

"Could I remember the recipe? No. Could I improvise? You have got to be joking." He thought of calling his mother, but remembered she was away on holiday. He was more comfortable with the computer than cooking, so perhaps in some reflex action he logged onto CompuServe.

"I'd never used my computer for anything useful in a real-world sense before," says Martyn. "Expecting disappointment, I was amazed to find the right recipe in the Cooks Online Forum [see Fig. 7-1]. I printed it, grabbed the sheet, ran back into the kitchen, and followed the recipe to the letter.

"The evening went smoothly and the Beef Wellington went down in history as the best we had tasted. The girlfriends didn't last, but the memories of the food and good times remain."

Paula Beasley, of San Juan, Puerto Rico, is a regular in the Cooks forum. "I love to cook, and I get bored with the same old stuff," she says, "so I like to find

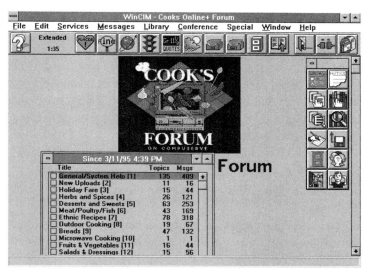

**7-1** *The Cooks Online Forum on CompuServe, with a partial listing of the discussion sections.*

out how to do new things. Last week I made my first bottle of flavored vinegar using a recipe I found on the forum. There are recipes in every category, and some really experienced cooks to help you figure out what, if anything, went wrong with yours. It's like having a cooking school at your fingertips.

"The same is true with gardening. I like to grow herbs and a few other things, and I can usually find a file in the [gardening forum] library to help me out, or an experienced gardener with a few words of advice. It's also a fun place to converse with people on topics not related to gardening."

Paula's last remark hits on half the reason people hang out in the cooking and gardening forums or subsections that are on every network: They have fun. For those of us who are urban workaholics and think the four basic food groups are restaurant, deli take-out, home-delivery, and microwave, it's hard to understand the endless fascination people in these forums seem to have with discussing their favorite recipes or restaurants or gardening tips. But the word *fun* comes up in talking with people about why they hang out there, as often as *food* or *flowers*.

Every time I lurk in the rec.food.cooking newsgroup on the Internet barely long enough to read the list of message headers, I invariably have to leave to get something to eat. Just the list of what people are discussing, alone, is mouth-watering. My response, of course, is to wander over the neighborhood yuppie deli and see what variation of chicken they've cooked up that day, but I admit to vague twinges and occasional vows to become a gourmet cook some-day. Sure. These folks are far more serious about that goal than I, but not so serious that they forget the fun part.

# What's online for cooks

There are enough online recipes, in countless libraries and databases, to last anyone several lifetimes. Here are a few good resources to start with on the Web:

*Amy Gale's Recipe Index* Indexed like a cookbook; the Web address is: http://www.vuw.ac.nz/who/Amy.Gale/recipes/&other-sites

*Cooking Index* Recipes from around the world: the Yahoo web address is: http://akebono.stanford.edu/yahoo/entertainment

Condé Nast's new Web showcase, called Epicurious, goes online in mid-1995. It will be home to articles and interactive features from some of their maga-zines, including *Bon Appetit* and *Gourmet*. Because they own Random House, the Random House cookbooks will be online there too. When it's available, you should be able to find it by using *epicurious* as a search term on the Web.

 If you don't have time to go shopping, there's a fairly broad selection of gourmet food gift packages, wines and spirits, and spices available to browse and order online, either in the shopping areas or the commercial services or on the Web, and more vendors set up shop weekly. (See Fig.7-2.) There are also grocery-shopping services and even a food co-op (see chapter 12, *100 ways to save time and money online*) where you can enter your order online and have it deliv-ered. Their geographical reach isn't very far yet, but this is one of the services almost every woman I've talked with mentions she'd like to have. The cus-

**7-2** *The Internet Wine Rack, one of several places to order wines and spirits online. This one's on the Web and has other interesting resources for connoisseurs.*

tomers are clearly there, so it's just a matter of time until the stores and co-ops become convinced of that and expand. For now, check Smart Food Co-Op on the Web, Shopper's Express on America Online, and Peapod, an independent service based in Evanston, Illinois, which is also available in some areas of California. Waiters on Wheels will also deliver meals from local restaurants. They're on the Web too, but currently serving only selected parts of the Pacific Northwest, although they're due to branch out.

To find culinary conviviality on the commercial services, check the leisure-interest or lifestyle areas, or use *cooking, food, wine,* or *beer* as keywords (home brewing is a big thing online, by the way). On the Net, to contact real people rather than Web displays and databases, check out the rec.food hierarchy of newsgroups. There are about half a dozen, covering vegetarian cooking, beverages, and restaurants.

# Garden tools

For gardening, use pretty much the same search strategy to explore: *garden, gardening,* or *home* will work as keywords to find both the forums on the commercial nets and the Web sites. The Internet newsgroups to look for here are under rec.gardens. Two good places to start on the Web are:

***Time Warner's Virtual Garden*** The Virtual Garden (Fig. 7-3) is located in Time Warner's Pathfinder service at: http://www.pathfinder.com/virtualgarden.html

***The Garden Encyclopedia*** This is in Books That Work at: http://www.btw.com/garden_archive/toc.html

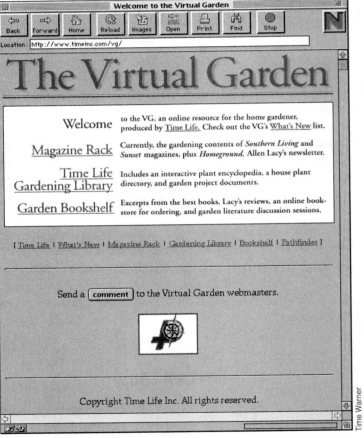

**7-3** *The Virtual Garden on Time Warner's Pathfinder service, on the Web.*

# Home maintenance

Imagine that it's a rainy Saturday afternoon, so your enthusiasm for going anywhere or doing anything is as damp as it is outside. Hearing the raindrops on the window reminds you of the leaky kitchen faucet you or your nearest-and-dearest have been promising for weeks to fix. You already bought the kit at the hardware, or maybe ordered it online and had it delivered. So, feeling put-upon but self-righteous about it, you tackle the dreaded duty. But it's not working. You can't get the whatchamacallit off, and the only friend or neighbor you know to call is away on vacation, yet paying the price a plumber would charge to fix a plain old leaky faucet seems ridiculous. Well, stop gnashing your teeth and log on to one of the home forums, where this very problem is one of the things people commonly ask about.

Day or night, you can always find people online who are so glad to be among the minority living the American dream of owning their own homes that they're nigh unto obsessed about anything and everything that goes with maintaining one, whether it's a co-op in Manhattan or a cottage in Montana. They'll patiently explain the difference between concrete and cement or tell you exactly how to get that horrid wallpaper off the wall of the charming old house you just bought without leaving bumpy blotches that show through the new paint or wallpaper, or simply how to fix that irritating leaky faucet. You can also shop for and buy the tools to fix the faucet or a whole new kitchen sink, for that matter. All online.

 You'll find authoritative advice on buying, building, or remodeling a home, plus abundant information on financing alternatives and free software that calculates mortgages. You can even apply for the mortgage online, find out how to fight city hall over a zoning ordinance, and hear what experts recommend about everything from cultivating prize roses to hiring a good contractor or making a bad one make things right. Being online not only makes tedious, time-consuming things easier, it makes fun things even more enjoyable, because you can share every stage of your saga with a group of people who look forward to hearing your updates and telling you theirs. All without leaving home.

You might wonder why anyone would go online and pay to talk about how to fix leaky faucets or installing drywall. The camaraderie is a big part of it. But they also do it because it's quicker and easier than going to the hardware store, or to the library or bookstore to get a book. If you already have the book, it's quicker and easier for you to use that, no question. But like cookbooks and gardening encyclopedias, home-repair and decorating books are gradually showing up online too, all updated, and all searchable to save you time. Ditto on the most popular home, garden, and food magazines, and they have more than just articles.

Perhaps even more useful than having forums where you can ask a quick question or dream along with other people who seem nearly as interested as you are in your remodeling project are the documents and software programs stored in the libraries. You'll find detailed diagrams about how to shore up your pier-and-beam foundation or build steps for a deck, charts on what foliage or ground cover does well in which part of the country, and sundry software programs for everything from an inventory of your household goods for insurance records—complete with date of purchase, cost, and serial numbers—to whether you're likely to get the loan for the house you want and how much the mortgage payments will be if you do.

Two shareware (try-before-you-buy) programs on GEnie are fairly typical of the software you'll find online. The first, called Grocer (filename GROCER34.ZIP), is one of many grocery-shopping databases, and a good one. It keeps a running list of your favorite products and their prices (some programs also sort by supermarket aisle and location), then adds the total dollar amount of your shopping list. You can also save your list so you don't have to reenter items you routinely buy. If you like leisurely meandering around the aisles at the store and if you have no budget constraints, then your ordinary handwritten list will do fine. But if you want to maximize your time and money, it's a handy tool. The second is a well-designed program to keep track of items and money at garage sales. It sounds like a much better idea than the usual chaos and handwritten stickers until you realize that to use the garage-sale program you'd have to use a laptop computer at the garage sale, much as if it were a cash register. That sounds not only like a real hassle, but a good way to get a laptop computer stolen. It's probably preferable to put up with chaos and stickers. But it doesn't take long to download either program, which means it doesn't cost much either (especially after 11 P.M., when downloads go faster), and it takes only a few minutes to try both programs and decide whether they're worth using.

It's also the library aspect of the magazines that makes them useful online in ways they aren't in print, even though most of us will always prefer to curl up in bed or on the sofa with a magazine we want to read cover to cover. Online, you can skim the current issue to see if there's anything you want to read, which saves trees and clutter, or search all back issues for a story on a particular topic, or find complete collections of features that appear in print only one per issue, such as the home plans from *Home* magazine on America Online (see Fig. 7-4).

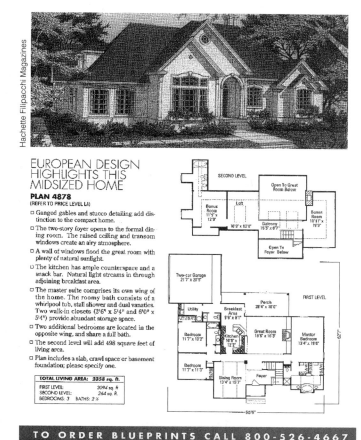

**7-4** *One of* Home *magazine's house plans from their forum on America Online.*

Hearst will premiere their new Home Arts site on the Web in the fall of 1995, with features and extra services about cooking, decorating, remodeling, and entertaining from their many magazines, including *Good Housekeeping*, *Colonial Homes*, *Victoria*, *Popular Mechanics*, *Smart Money*, and *House Beautiful*. In addition to the articles, Home Arts will include many interactive features. For instance, you can get advice from an expert on buying, selling, or refinishing antiques; calculate how much paint you need to buy for the living room just by entering the dimensions; find out what to do about the red wine you just spilled on your rug; or get suggestions about what you can make for dinner when all you have in the refrigerator are a chicken, an avocado, and ketchup. There are similar services scattered elsewhere online that do some of that, and software that does the rest (with demonstration versions available online, of course), but what Hearst hopes to create is a one-stop-shopping place that will lure, as spokeswoman Diana Dowling puts it, "people who care more about their homes than their computers." Use the keywords *home arts* to find it on the Web.

# The revolution in real estate

 Online databases are changing the way real estate brokers work forever. No longer will you be dependent on brokers just because they have access to the Multiple Listing Service database and you don't, because you will too. It will be online, so you can browse it from home without sales pressure, and contact the broker only when you've found a house you want to see. Many people will list their property for sale or rent directly, too, with one of the many real estate services or classified ad sections online. The listings on the commercial nets are still sparse, and the ones on the Web aren't a lot better, but they will be. The best ones will have photos of the houses and demographic data on the neighborhoods, and allow you to search by city or zip code, as some already do.

One of the best so far is Home Buyer's Fair, on the Web (see Fig. 7-5), which provides links to other home and real estate resources and a quite comprehensive set of tools for anyone shopping for a home to buy or rent (although, as with offline realtors, there aren't many rental listings). There's a mortgage qualification calculator with a tipsheet, lists of lenders by state and easy ways to

*Welcome to the Homebuyer's Fair*

**7-5** *Home Buyer's Fair, on the Web at http://homefair.come/homegrap.html.*

send them inquiry postcards directly from Home Buyer's Fair, guidelines on how to shop for a mortgage agreement, a weekly interest-rate survey, classified ads, and more. Most of the real estate sites on the Web are still new, so give them time to expand their listings as realtors, sellers, and landlords see the value of being online.

There are also classified real estate ads on the commercial services, and sections for homeowners or those who hope to be someday. Just use *real estate*, *house*, or *home* as a keyword, follow the signs, and don't lose sight of the dream.

# CHAPTER 8

# In sickness and in health

**L**ike many women, in her first foray online Susan Heinlein's impression was that it took too much time and too much money. So she eventually canceled her CompuServe account without exploring much. That was before a doctor told her long-time partner, Merrill, that he had lung cancer, and had less than a year to live.

"Suddenly, I was desperate for information, any information," says Susan. "And I felt so alone."

After a friend sent her printouts of what she'd gleaned from the National Cancer Institute's CancerNet files on the Internet, Susan put aside her skepticism and signed up again, this time both for CompuServe and for Netcom, her nearest Internet service provider. What she found surprised her, and was far more than she'd expected.

*The people in the Cancer Forum were right there for me,*
*24 hours a day. It was just amazing.*

From the Internet, Susan downloaded more information from the NCI, and from the FAQ (frequently asked questions) files from the NIH and the American Cancer Society. She also signed up for a newsgroup called alt.suppport.cancer.

"But I found that the Internet didn't really fulfill the emotional need that I had, even in the newsgroups. I was so frazzled. I wanted to just talk with anyone. My needs were so little, yet so great."

So she joined CompuServe's Cancer Forum, which someone on the Net suggested she might like better than the newsgroups because forums on commercial services are more structured. She used both, but to fill different needs.

"The people on CompuServe, from all walks of life, were right there for me, 24 hours a day. Merrill's doctor was kind of cold and didn't like me asking questions, and my friends didn't know anything. So I had cognitive questions and emotional questions. The people in the forum said, 'Here are the questions you need to ask and the places you need to go.'

"At any given time, the Cancer Forum has six to eight active oncologists, plus researchers from schools of medicine, pharmacologists, academics, and other clinicians. There are also patients and caregivers like myself, and just people who have been there. When I first came online, I introduced myself, and told what I knew about Merrill's situation. Fifteen people, three or four of whom were doctors, responded immediately.

"Physicians asked specific details about the diagnosis and prognosis, and would immediately get back with me. The previous patients were reaching out to me and giving me cyberhugs constantly. Caregivers offered pointers on taking care of myself. One of the physicians even went so far on two occasions to talk with Merrill's doctor, long-distance, for lengthy periods of time. It was just amazing.

"It was even more than I could have expected face to face, because a lot of my friends and acquaintances tended to back away. Or they'd say, 'I know what you're going through,' and I felt, 'No, you don't.' I needed to be with people who truly did know. Yet I'm 30 minutes from the nearest town, and I didn't want to give up my time with him to be bouncing around to local support groups."

*If I ran into a crisis that I could put a name to, the first thing I would do today is to go online and ask questions.*

There were no magic cures available online, of course, any more than offline. So just four months later, Merrill died. But Susan continued to get support from the people in Cancer Forum in dealing with her grief. And losing Merrill, who was only 53, made her rethink what was important to her. She's a freelance writer, so she volunteered to write for a new lobbying group for cancer patients, and helped create and edit an anthology of essays, poetry, and short fiction on grief from women throughout the country.

Susan's experience also changed her mind about the value of investing time and money in online services. "If I ran into a crisis that I could put a name to, the first thing I would do today is to go online and ask questions. That's very different from what I would have done five years ago."

# From mysterious symptoms
# to useful solutions

The friend and neighbor who prodded Susan to get online is Carola Draxler, who learned the value of online services first-hand when she had to cope with a health problem of her own.

Carola was a partner in an electronics service and consulting business, which was going very well. They had clients throughout the San Francisco Bay area and she commuted over winding, two-lane mountain roads to the office she shared with her partner. But that was normal. Because of the nature of her work, she spent long hours at her computer, which meant she had to make sure she kept mobile enough to flex and stretch her muscles, lest she develop wrist, arm, and back problems. But that's normal in this high-tech era too. To counter the effects of being tethered too often to the computer keyboard, she frequently took long walks along the steep sideroads and among the redwoods near her house in Boulder Creek, California.

*It was very useful in the beginning. It was really
like a sit-down kind of support group.*

But Carola gradually noticed that something wasn't normal. Despite that conscientious exercise, she sometimes had trouble walking.

"Then I started feeling burning or stabbing sensations, and generalized pain," she says. "It's like when you have the flu and your body aches all over, but three times the intensity of that. It became more and more debilitating." So debilitating that she had to dissolve the partnership for her consulting practice, partly because she could no longer work full-time and partly because of problems that had developed between her and her partner because of the continual strain Carola was under.

She went to a rheumatologist who suspected that fibromyalgia could be the cause of her symptoms, but he wanted to consider other possibilities too. "Fibromyalgia isn't something a lot of doctors know a lot about," says Carola.

"The majority of the sufferers are women, so until fairly recently, women would go to docs with this laundry list of symptoms, and be told to go home, take a warm bath and reduce stress, or they were sent to psychologists. So it's not uncommon, but it was being dismissed as a 'woman's thing.'"

While she was waiting for her doctor to tell her what was causing her problems, the pain and uncertainty were driving her bonkers, so Carola kept looking for answers on her own. "I was reading the Science-Med newsgroup on the Net, then an arthritis newsgroup started and there was some mention of fibromyalgia in that. Because my doctor had mentioned that term and I was following every trail, I read a bit about fibromyalgia and thought, 'That sounds exactly like what I have.'

"That's what my doctor finally diagnosed. So my pain finally had a name, which meant I could look for solutions. Next, I heard online about a mailing list (listserv) called fibrom-l, so I subscribed to that, and was just flooded with posts. It was like reading about *me*. I learned a lot from the people in that group."

It was because of what she learned from that fibromyalgia mailing list on the Net that Carola began taking a sleep aid sold in health food stores, with her doctor's approval and in addition to the medication he prescribed. It was the reports from others about how much aerobic exercise had helped them that prompted her to buy a treadmill so she could resume walking, at least in some fashion. And from a newsletter she subscribed to after learning about it online, she found a fibromyalgia support group in a town within easy driving distance. When she couldn't get there, she logged into the real-time fibromyalgia support group hosted by the Better Health and Medical Forum on America Online, on Sunday evenings.

"It was very useful in the beginning. It was really like a sit-down kind of support group, in the sense that I compared notes with people who had it, and found out what works for them and what doesn't. It was encouraging, because I learned that you can go into remission.

"Now I might poke my nose in once a week, but I don't read it very much anymore," she says, "because I can't have my whole life be that. I'm not interested in just going there to kavetch. I find it more fun than helpful, because I have to wade through an awful lot of text from other people to get anything new anymore. Information still comes in pieces when I least expect it, but I think my information-gathering stage is finished."

# How to find help and information

As you can see by Susan's and Carola's experiences, there's more online than just information, and it's not just the information that makes the information superhighway super. It's the combination of moral support and information that people give so freely to each other, and how you use that information in real life—all without leaving home—that makes online services truly useful. In chapter 18, *Online research basics*, you'll find a detailed guide on how to search some of the best databases on the commercial services, using fibromyalgia as an example. That chapter also compares costs, with surprising results. Here are some highlights of what's where in health information and support.

– 120 –

# The best places to start looking

Of the commercial networks, vs. the Internet, CompuServe has by far the most research resources. Except for the extensive government archives, you can find just as much helpful information there as you can on the Internet, and you can find it a lot more easily. It will cost more, true, but it's still a bargain when you're worried about your health or the health of a loved one. And if your need for information is urgent, the last thing you probably feel like doing is roaming the Net. So start with these, and if you want more go to the Internet.

## HealthNet

(On CompuServe: FIND health or GO healthnet or GO reference) If all you need is rather general information about a relatively common disease, HealthNet is a good place to start because it's almost free. That's because CompuServe classifies HealthNet as one of their basic services, so it's covered by your monthly fee. Once there, your initial choices are The HealthNet Reference Library and Sports Medicine. Choose the first, and you'll get the following menu of choices, which is similar to the menus you'll see in any searchable, text-only online database:

```
HEALTHNET REFERENCE LIBRARY

1 Introduction
2 Disorders and Diseases
3 Symptoms
4 Drugs
5 Surgeries/Tests/Procedures
6 Home Care and First Aid
7 Obstetrics/Reproductive Medicine
8 Ophthalmology/Eye Care
```

Each of these choices leads you to another list, and eventually to short descriptions or documents, based on whatever options you choose in the previous step. Unfortunately, there's nothing about fibromyalgia under Disorders and Diseases or any of the subcategories, which include Psychiatry, Nerves/Brain, Skin Diseases, Breathing/Lungs, Heart/Circulation, Digestion/Liver/Stomach, Urinary, Cancer, Infections, Gynecology/Female Organs, Hormonal/Endocrine, Arthritis, and AIDS/HIV. Nor does it show up in HealthNet's Index to Diseases. HealthNet might help for something else, but obviously not for this topic. Nonetheless, it's good to know what you can find there some other time, because it's the only health reference database on CompuServe that isn't surcharged.

# Health Database Plus

(On CompuServe: GO HLTDB or GO healthDB) Health Database Plus includes online versions of consumer magazines, such as *Men's Health*, *Parents Magazine*, *Prevention*, and *Runner's World*, as well as reports and journals from institutions such as the Centers for Disease Control, pamphlets by the American Lung Association and other organizations, and professional journals, including *JAMA*, the *Journal of the American Medical Association*, *The Lancet*, and *The New England Journal of Medicine*. Only abstracts are available for most of the journals, but you can get the full text of most magazine articles. It also incorporates relevant full-text articles from publications that aren't specifically about health. Health Database Plus is updated weekly, but because of the limitations of ASCII text, they can't include graphics like bar and pie charts, although you will get the raw data from most tables. You pay only regular rates for search time, but any full-text articles you download are $1.50 each. For helpful tips and more details, see chapter 18, *Online research basics*.

# Knowledge Index

(On CompuServe: GO KI) At $24 an hour, or 40 cents a minute, this one costs considerably more, but you get much more, including access to Medline, which contains the archives of the National Library of Medicine. Because of its depth and breadth, if you learn how it works and a few tricks, Knowledge Index can actually be the most cost-effective, as you'll see in chapter 18. There are five key steps to ensure that you find the most information, yet spend the least money:

1  Download all the descriptions and instructions first (that part doesn't cost extra), read through them offline and decide which database sounds like the right one to use.

2  Read through the instructions about how to define and refine your search terms, then write them down on paper, keeping them as specific and as narrow as possible.

3  Go online and set your capture buffer to save the whole search to a file in your computer (often the PageDown key toggles it on and off, depending on your software, or look on the menu bar while you're online for "capture"). Do the search, tag the citations as quickly as you can, and log off.

4  Read through the citations offline, and choose just the most important ones. You'll be much more judicious about this when you can see the entire list of what you found in step 3 in front of you, offline, than you will if you ignore this process and tag what you want while you're online.

5  Have your list highlighted and handy, and go back online to grab the articles you want, log off again, and read through them leisurely, offline.

## IQuest's Medicine Database

(On CompuServe: GO IQ —> medicine or GO Paperchase) This, too, will get you to Medline, only here they call it Paperchase. However, the way it works is so complicated and the surcharges are so high compared with the other services, that it's not for novices and not the route for anyone to go unless someone else is paying for it and cost isn't a consideration. There are exceptions and ways to get around that, so see chapter 18 if you'd like to know more.

For prescription drug information, forget trying to understand the *Physician's Desk Reference.* There are three excellent sources of information about prescription drugs at your fingertips:

## Consumer Reports' Complete Drug Reference

On CompuServe, (type GO drugs) One pharmacist says, "This is essentially the consumer volume of the United States Pharmacopeia's dispensing information and, hence, is very authoritative."

## Knowledge Index's Drug Information

From the American Hospital Formulary Service's *Drug Information and Handbook on Injectable Drugs.* It also has a category called Unlisted Drugs.

## PharmInfoNet, on the World Wide Web

This one is more clinical because it's also for medical professionals, but it's intended for patients too. The Web URL address is http://pharminfo.com/.

As Susan Heinlein says, "Women ought to do better at finding things faster online than men, because men have more problems with asking for directions than women do. That's what it really is all about—asking 'how do you get to such-and-such?'"

# Healthy living

Lest you think that all the health information online is disease-related, rest assured that that's not so. There's an abundance of information on nutrition, as well as quite a lot on fitness and exercise, and plenty of places for those who don't spend all their time staring at computer screens to hang out too. Beth Jordan, a secretary in Dayton, Ohio, is more into sports and fitness than most of us, and holds a racing license in one of the women's categories of the United States Cycling Federation. She says she prefers services like CompuServe, where she's a regular in the cycling section of the Outdoor Activities Forum, to the Internet newsgroups, because it's easier to feel part of a community.

*It's like a big family in the cycling section of the forum. We know what kind of bikes everybody has and their kids' names.*

"The Internet is like a vast wasteland, because it's so big and so complicated," says Beth, "but it's like a big family in the cycling section of the forum. We know what kind of bikes everybody has and their kids' names. A lot of the people are rather techno-geek oriented, and that kind of person gets into fiddling with bikes. I don't mind tearing down my bike and doing work on it or installing a modem in my computer, for instance. Most cyclists are pretty social people; they like to get together and tell bike stories. So it's like a global bike shop where you go in the back room, hang around with the mechanic, and shoot the bull. There's a competition section for people like me who are heavily involved with racing, a touring section for those who want to pack their bikes up and go someplace relaxing for the weekend, a general section for people who just like to tell stories or hang out, and a maintenance or equipment section for how to find X bolts or Y gear sprocket. I've learned a lot from people there, and they were a big factor in pushing me over the edge to get my racing license."

To find out more about the active sports areas, see chapter 10, *Pleasurable pursuits.*

# Highlights: Online health discussion areas

Without any authoritative evidence at all to back this up, my guess is that if anyone analyzed the content of what's available online, you'd find health information among the top six categories, along with information about computers, genealogy, small businesses, entertainment, and personal finance. So a complete rundown would require a whole book. As a case in point, here's what you get just on CompuServe if you type FIND health at the FIND icon:

```
Place (+ means Surcharged area)  GO word
  1 Aids News Clips +             AIDSNEWS
  2 Audio Book Club(FREE)         AB
  3 CCML AIDS Articles($)         CCMLAIDS
  4 CNN Forum +                   CNNFORUM
  5 Cancer Forum +                CANCER
  6 Consumer Reports Drug Ref.    DRUGS
  7 Contact Lens Supply(FREE)     CL
  8 Diabetes Forum +              DIABETES
  9 Disabilities Forum +          DISABILITIES
 10 FORTUNE                       FORTUNE
```

```
11 Florida Today Forum +           FLATODAY
12 General Computing Forum +       GENCOM
13 Government Giveaways Forum +    INFOUSA
14 Handicapped User's Data         HANDICAPPED
15 HarperCollins Online(FREE)      HARPER
16 Health & Fitness Forum +        GOODHEALTH
17 Health & Vitamin Express(FREE)  HVE
18 Health Database Plus($)         HLTDB
19 Health/Fitness +                FITNESS
20 HealthNet                       HNT
21 Holistic Health Forum +         HOLISTIC
22 Human Sexuality Databank +      HUMAN
23 IBM Clinton Health Plan +       IBMHEALTH
24 IQuest($)                       IQUEST
25 Lens Express(FREE)              LENS
26 Magazine Database Plus($)       MAGDB
27 Medsig Forum +                  MEDSIG
28 New Age Forum +                 NEWAGE
29 PaperChase-MEDLINE($)           PAPERCHASE
30 Physicians Data Query($)        PDQ
31 Rare Disease Database +         NORD
32 Retirement Living Forum +       RETIREMENT
33 Sundown Vitamins(FREE)          SDV
34 Syndicated Columns              COLUMNS
35 The IAMS Company(FREE)          IAMS (pet food)
36 Time Warner Lifestyles Forum +  TWLIFE
37 U.S. News & World Report        USNEWS
38 Vegetarian Forum +              VEGETARIAN
```

What's more, although CompuServe's keyword search system is better than most, as is the case with every online service's keyword function, this one misses several valuable places, such as the ADD (attention deficit disorder) Forum, the Multiple Sclerosis Forum, the Muscular Dystrophy Association Forum, and the health sections in the Women's Forum. So the availability of information and support isn't a problem, but finding it and then the range of choices can be. I've yet to see a book or magazine article that caught everything, and things change constantly anyway.

Thus the best advice I can give you is to use very specific keywords to search in the main directories, then if you don't find what you're looking for go to one of the general health forums and ask people there if there's a place on the topic you want to find. If you're not looking for something as specialized as the Multiple Sclerosis Forum, check the sections in the various general health forums to see what appeals to you. Even then, each forum develops a character of its own. So the sometimes new-agey slant of the Health and Fitness Forum on CompuServe, for instance, might not appeal to you, or it may be just what you'd hoped to find. Only you can decide what's for you, but these guidelines will help.

# Recommendations on the commercial nets

If you disregard CompuServe's research databases and specialty forums, and look only at who has the forums with the greatest range of general-interest topics and activity, including weekly real-time support groups, America Online comes out on top, with CompuServe and Prodigy second, then Women's Wire, which is a contender more because of its focus on women's health concerns than for its level of participation at this point. But Women's Wire is also newer and much smaller than the other two, so the fact that it's a contender counts for something. For specific diseases, such as cancer, diabetes, muscular dystrophy, and multiple sclerosis, CompuServe takes top honors, hands down.

## America Online

For all-around usefulness, the Better Health and Medical Forum on America Online (see Fig. 8-1) is pretty comprehensive. Sample sections are Lifestyles and Wellness; Mental Health and Addictions; Human Sexuality; Informed Decisions; Health Reform and Insurance; Men's, Women's, and Infant and Children's Health; Seniors' Health and Caregiving; Alternative Medicine; and Home Medical Guide. They should get bonus points for the many support groups they run each week as real-time chats within the forum. Among those are Chronic Illnesses and Conditions; Living with Cancer; PC Pals' Deaf Teen

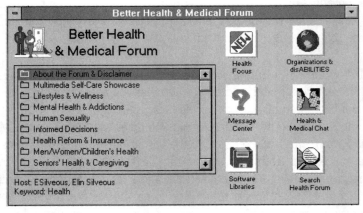

**8-1** *America Online's Better Health and Medical Forum.*

Chat, sponsored by Alexander Graham Bell Association for the Deaf; Deaf/Hard-of-Hearing Chat; Fibromyalgia; General Disabilities; Chiropractors' Professional Networking; Endometriosis; Multiple Sclerosis; Occupational Therapists' Networking; Loved Ones of Cancer Survivors; Down Syndrome; Post Polio Survivors'; Herpes; Repetitive Strain Injury Self-Help; Cerebral Palsy; Chronic Conditions; Multiple Sclerosis; Nurses' Networking; Muscular Dystrophy; and Disabilities Funtime.

## CompuServe

The Medsig Forum on CompuServe is geared toward physicians, nurses, and other health professionals, although they welcome laypeople who have questions too. They don't dispense medical advice, of course, because that would be unethical, but they're good about guiding you in the right direction, and are usually quite empathetic if you come there in the midst of your own or a loved one's health crisis. As for the Health and Fitness Forum, sometimes referred to as the Good Health Forum, here's the lineup: General/Help, Healthy Recovery, Mental Health, Family Health, Doctor's Inn, Exercise and Fitness, Running and Racing, Nutrition, Martial Arts, Health Networking, Women's Health, CFS/CFIDS/ME/FMS, and Self Help/Support. The Holistic Health Forum has a Women's Health section, and an excellent roster of others if you're either curious or already serious about things like vitamins, herbs, and yoga. Also see the directory earlier in this chapter, and check into the Women's Forum, Lifestyle Forum, and, if that's your bent, the New Age Forum.

– 127 –

## Prodigy

Prodigy's Health and Lifestyles Bulletin Board topics include: Bodybuilding, Cosmetic Enhancement, Dental Care, Diet/Nutrition, Emotional Therapy, Exercise, Eye Care, Family Medicine, Foot Care, Health In Media, Holistic Medicine, Home Remedies, Hygiene, Injuries/Healing, Internal Medicine, Men's Health, Over the Counter, Pregnancy, Safety/Prevention, Skin Care, Sleep/Dreams, Stress Management, Women's Health, Vitamins, Weight Control, and Other.

## Women's Wire

Health & Fitness has an excellent framework and focus, and can only get better as Wire grows and has more participants. Key areas are: Pregnancy & Birth, including Childbirth Preparation; Aging, which includes Alzheimer's Disease, Aging & Longevity, and Caring for Elder Parents; and Women's Health Issues, including Women & Cancer, Women & AIDS, Women & Weight, Women & Alcohol, Menopause, Chronic Fatigue Syndrome, D.E.S., and Breast Implants. Women's Wire also subscribes to the Women's Health List on the Internet, so you can browse it and read what you want without having all of it dumped into your e-mail box daily. They also have files of information with lists of health information hotlines and other resources.

## Worth mentioning

On GEnie, the best place to check is the Family Roundtable, as well as the private women's section within that. They also have a Medical Roundtable that's more for doctors. On Delphi, look for the custom forums called Today's Woman, Living Healthy, Homeopathic/Holistic Health, and Child Health Forum.

## Health resources on the Internet and the Web

Some of the very best information and support on the Internet is available through the mailing lists, because they're more private and enable people to be more candid. They also stay much more focused than most of the newsgroups. So study the List of Lists or the List of Publicly Accessible Mailing Lists if you're looking for help or information on a specific health concern. Most likely, there's a list for it, and anyone can subscribe. One way to find out is to join the HMatrix-L mailing list, which is a discussion of online health resources of all kinds, carried on through e-mail. You can cancel, or "unsubscribe" in Internet parlance, whenever you want. To subscribe, send e-mail to: listserv@kumchttp.mc.ukans.edu and in the body of the message, type SUBSCRIBE HMatrix-L followed by your real name. For example, SUBSCRIBE HMatrix-L Jane Q. User.

The Yahoo and WebCrawler search engines are also great ways to narrow down your choices, and with these, unlike the List of Lists, you don't have to read through hundreds of pages because they're sorted by subject or category, plus you can do a keyword search. But if you want just one source to look through for references to health information on the Net, the best one is the Subject-Oriented Clearinghouse at the University of Michigan. Within the Health category, there's a subcategory for Women's Health, so be sure to check that too. This will show you how to get to the Columbia University/Barnard collections, which are definitely worth perusing. To browse for newsgroups that might interest you, start with alt.support, soc.health, and misc.fitness. If you agree with Women's Health Action and Mobilization, a.k.a. WHAM, that "women's health is political," then try these two mailing lists:

➤ To subscribe to the WHAM list, send the message SUBSCRIBE WHAM *yourfullname* to listproc@listproc.net.

➤ To subscribe to the Healthy Cities Women's Network mailing list, called CITNET-W, send a similar message with SUBSCRIBE and your full name to listserv@indycms.iupui.edu.

## AIDS

The following is the AIDS Information site on the Web: http://cornelius. ucsf.edu/~troyer/vanews/. At the same site, look for The Safer Sex Guide: http://cornelius.ucsf.edu/~troyer/safesex.html.

## Cancer

OncoLink, at the University of Pennsylvania, is the best starting point on the Web for cancer information and support. It will lead you to information about other Web sites, newsgroups, mailing lists, and databases throughout the Internet. There are also good pointers to several support groups. The Web address is http://cancer.med.upenn.edu/. Another outstanding site is the Breast Cancer Information Center by NYSERNET, the New York State Education and Research Network, shown in Fig. 8-2. Its Web address is http://nysernet.org.bcic/. Both of these are exceptionally good sites, and will make at least the information-gathering stage of coping with cancer a little easier.

## Diabetes

The best route is to go to the University of Wisconsin Med School's Diabetes Knowledgebase, at: http://islet.medsch.wisc.edu/ —> Diabetes Center.

Maintained by the *New York State Education and Research Network*

**8-2** *Breast Cancer Information Clearinghouse.*

## Directories

Another good where-to-find-it resource is the Guide to Internet Medical Resources sponsored by the American Medical Informatics Association, the same organization that sponsors the Medsig Forum on CompuServe. Like the University of Michigan's Subject-Oriented guide, this is a good place to start if you just want to find out what's online, then choose areas to visit. The Web address is: http://kuhttp.cc.ukans.edu/cwis/units/medcntr/Lee/HOME-PAGE.HTML

## World Health Organization

Finally, if you want the whole picture, go to Geneva, Switzerland and take a tour of the WHO site on the Web at http://www.who.ch/. All from your computer keyboard, of course.

# CHAPTER 9

# Lifelong learning

In no other realm is the ability to communicate by modem going to have a more profound impact than in how and where we learn. It's one of the most exciting ways online and related electronic media are being used, and one that's going to make a whole generation of people who don't understand it feel threatened.

# Colleges and universities

"We're moving towards a society in which education will be a lifelong enterprise," says Stephen Anspacher, Director of Distance Learning for the New School for Social Research in New York City. "People will continue to go to colleges for the first two years of an undergraduate program, then finish their baccalaureates online. Any subsequent education will be based on need, using a variety of educational resources. Online education, videoconferencing, TV, phones, and computers will bring people an opportunity to take courses and get degrees from schools that are good in whatever field they're interested in, and they'll be able to put together a program from the best schools from a whole menu of courses. . . . By ten years from now, an increasing amount of undergraduate, professional, and continuing education will be done this way."

The New School has always been in the vanguard of innovative education, so they already have a site on the World Wide Web to publicize their online curriculum (http://www.dialnsa.edu/home.html), as well as a way for people to request information by e-mail (info@dialnsa.edu). They offer several courses and a master's in media studies online.

*We're moving towards a society in which
education will be a lifelong enterprise.*

It's not just the existence of multimedia technology that's bringing this shift about, Anspacher says. "The pressure is growing as tuition becomes increasingly expensive. Plus there's the lost income and expense of living while going to school, and people want to achieve a balanced life without a lot of juggling of their schedules."

Over the next 50 years, Anspacher predicts, "The 3,500 schools in existence in the U.S. now will merge and emphasize their strengths, just as we've seen happen in corporate America. Those still here will be offering courses both on campus and at a distance. We'll have half the number of higher education institutions in the country, but the smaller number will be much richer. . . . In the beginning, it will cost about the same, because the business side of academe hasn't figured out how to market themselves in this way." Eventually, theoretically, education will cost considerably less.

# Elementary and high schools

Even though Anspacher says it will take a generation for all of this change to occur in higher education, it's already beginning to happen in elementary and high schools. Kids have taken to technology enthusiastically all along, and now teachers who were skeptical and scared are getting excited. The Kidsphere newsgroup for K-12 teachers and students on the Internet is one of the busiest anywhere. Just as companies are creating project teams with people in different countries working together through various forms of electronic communication, teachers and students of all ages are establishing relationships with classrooms in other countries, doing projects together, and replacing pen-and-paper communication between foreign pen pals with e-mail correspondence. The Edupage mailing list, which is about creative ways to use technology in education, is another very active network. All this is not just about old ideas and new technology. Many take the initiative to connect informally with like-minded teachers through a newsgroup. In other instances, it's more of a coordinated, official effort:

*The National Geographic Kids Network* Prepares different science- and geography-oriented teaching modules, then pairs students with other students at different schools around the world to study phenomena such as acid rain. The kids get to use the tools that a scientist would, and the software that comes with this National Geographic program provides a way to graph the results. Then a National Geographic scientist helps them interpret their findings.

**The School Web Exploration Project** Matches corporations and community organizations that have the computer capabilities to create a Web page with classes and schools who want to learn how to publish information online or collaborate on projects. Some projects are strictly local, and some connect students in different countries so that cultural exchange is a by-product. You can visit the School Web home page at the following address: http://k12.cnidr.org:90/swep.html.

What we need to think about, long-range, is what children learn, not so much how they learn it, says Therese Mageau, editor-in-chief of Scholastic's *Electronic Learning* magazine. "The problem is not what technology we should use, but what our children should be learning and be able to do when they leave school. People in their mid-30s and beyond were the last generation well-served by the American public school system, because it was designed for a mass-production economy. Now we have an economy based on information and how to get it and understand it, then make something out of it and turn it into something that's profitable. That requires very different skills than mass-production skills. So it's very hard for people in that transition generation to take that imaginative leap and understand."

A good illustration of how little many parent-age people understand what's already happening and what's coming was a rather irresponsible editorial in the *Cleveland Plain Dealer* in January 1995. It railed against the State of Ohio for spending money to wire classrooms for modems or even to buy computers, because all the equipment would just have to be updated or replaced frequently and that would be too expensive for taxpayers. "Instead, let's send business a whole generation that can read and understand the classics, write down a well-developed and well-informed thought, and do algebra without a calculator," wrote Kevin O'Brien, deputy director of *The Plain Dealer*'s editorial pages.

"Nobody does algebra without a calculator anymore, not even professional mathematicians," was Mageau's response when I faxed her that column. "To waste 200 hours of a child's life to make him into a poor imitation of a five-dollar machine makes no sense. Sure, we all need to be able to do certain kinds of mental math. When you're in the grocery store, you have to be able to know if you're putting more things in your cart than you can buy with the money in your wallet. But you don't need to be able to do long division in your head. The reason you need that number from long division is more important than how you get that number. The application of the knowledge is really what's key here. Except for people in service-sector jobs—cleaning floors and making beds—the kinds of skills being asked of them are not the kind of skills being taught.

"Ten years ago, the manual that an average car mechanic needed to do the job was a couple of hundred pages; now it's the equivalent of ten New York City phone books. You need more than good reading skills to handle that job. That car mechanic cannot do his job without a complete level of comfort with technological tools. The phone repair person who comes out when we have a problem doesn't climb a telephone pole; he sits down at a computer. Even the person who climbs the poles has a computer hanging off his belt.

"Most vocational education programs are very high-tech now. If you want to work at Pizza Hut, you've got to know how to work a computerized oven; if you want to work at Sears, you need to run a computerized cash register. The issue is: do we want our children to have a comfort level and feel in command, or feel completely lost?"

As Anspacher said about the reasons behind the growth of distance learning in higher education, Mageau says the impetus for the changes affecting school-age children aren't all idealism either, and some of this evolution is influenced by trends in the parents' work lives. "Most American families now need two incomes. There's going to be a tremendous movement to deregulate schools, so I think there will be opportunities for parents to take the $5,000 that their state spends to educate a child and spend it in the way that they want. So maybe the child goes to school for half the day, then goes to a museum school for a couple of hours, then works with the parent at home for two hours, because that parent is telecommuting part of the time. I think it's going to fuel home learning tremendously."

Some home-schooling advocates would have you believe that all children will be taught at home eventually, but that's not going to happen, says Mageau, because parents won't let it happen. Not all parents want to or can afford to stay home, and not everyone wants to or can telecommute.

# Adult education

As Anspacher pointed out, it's not just school-age children whose lives are changing. We're all going to be expected to take more courses just to keep current in our fields and, as it becomes easier and less expensive to take courses just for personal interest, more of us will want to, just for the challenge and fun. On GEnie and America Online, you can take academic-type or leisure-interest courses for as little as $25 for 12 weeks, all without leaving home.

"The vision that's being offered right now is learning anytime, anywhere," says Mageau. "Technology frees up learning so that it doesn't have to be in a classroom. It can be in your car, your home, your office, or at the local library or local museum. When you free information from physical constraints, that frees the learner. Sometimes it's called just-in-time learning [or learning on demand] because it's like 'I need to know how to do this particular thing right away, so teach me now.'"

Dana Willhoit, of Richmond, Texas, says it's only because she can take courses online for a master's in journalism that she's able to work toward the degree at all right now. "I have a six-year-old and an eight-month-old baby," she says, "and with a baby it's really hard for me to be away from home for long periods of time. Just the commuting time would be inconvenient. It would probably be years before I could get a master's degree any other way."

*The vision that's being offered right now*
*is learning anytime, anywhere. Technology frees*
*up learning so that it doesn't have to be in a classroom.*

The six-week courses Dana is taking are offered by the University of Memphis through a section of the Journalism Forum on CompuServe. "It's pretty intensive," she says. "There's a lot of reading, and papers to write, and the class meets online once a week for three hours. Part of your grade depends on participation." The $1,000 course is too expensive for many people; fortunately, fees for other online courses offered by universities are typically much less.

# What the future holds

## George Lucas' Edutopia project

George Lucas has already made his reputation as a visionary in film, and now he's applying his imagination, as well as his considerable financial resources and influence, to what he envisions education being like in the year 2020. By 1996, the Edutopia project, sponsored by the George Lucas Educational Foundation, will release a fictional film that portrays that vision and a book that documents how it can be achieved, with examples of actual programs already underway.

"What's missing from the national debate about how to change the system is a vision of what it could be like," says Patty Burness, executive director of the foundation." So that's what the Edutopia project plans to provide. "Our target audience is corporate executives, opinion leaders, policy makers, the media, and parents. It's not the educators because they cannot change the system alone. They know that and we know that."

Burness talks a lot about project-based learning, in which students all over the country and throughout the world will work together, using all forms of electronic communication, collectively known as *multimedia*. "Text, graphics, sound, and animation bring learning alive, and allow people to learn according to the way that really meets their styles of learning," she says. "You can get students and teachers solving environmental problems, solving social problems—whatever it is, they can respond and be a part of the solution. It allows people to develop relationships they wouldn't have been able to develop before, not only because of physical distance but because of the other biases we bring to face-to-face communication.

"We view the opportunity for the technology to free up the teacher to spend more time one-on-one in small groups with kids, so that they don't have to be responsible for 30 kids at a time. We envision lots of adults in the classroom. It's not going to be just a teacher who teachers math. It's an interdisciplinary world, where the teacher becomes a guide on the side rather than a sage on the stage. What children are learning today is too abstract, and they can't make the connection between that and the real world. Doing projects is a great way for children to deal with things in an interdisciplinary way."

Burness likes to quote what George Lucas often says: "Education is the core of everything." Lucas hopes to provide the vision, and persuade the politicians to provide the means to make it a reality. Says Burness: "When John Kennedy said we were going to the moon, we did. We had a vision, a strategy, a budget, and we got there."

## One woman with the courage and conviction to be a catalyst

On Women's Wire, I met Claudia Lamoreaux, an educational designer, multimedia artist-producer, and writer, who was unaware of the Edutopia project and doesn't have anything like that kind of financial backing, yet has a similar vision and the courage to be a catalyst. She organized a series of dialogues that began on Earth Day 1995, in San Francisco, which she'll make available through Haven.Net, her own Web site.

*The new technologies are revolutionizing education right before our very eyes.*

"The goal of these dialogues is to envision what 21st-century education could look like, and to offer models that can inspire people to imagine and create new-paradigm learning environments," says Claudia. "The dialogues are transcribed and videotaped, and key segments will be posted to the Internet Web site of Haven.Net. At the same time, questions will be posted and responses will be shared in the dialogues, so there will be a living interplay between the San Francisco dialogues and the global community. The final results will be put into a book about vision of the future of education."

Claudia describes education as her life's passion and has been involved in it for several years, sometimes as a writer or editor, often as a one-woman crusade. "The new technologies are revolutionizing education right before our very eyes," she says. "It's already like a parallel universe developing alongside the schools, because they're preparing young people for a world that already no longer exists. . . . Read *School's Out* by Lewis J. Perelman (1992, William Morrow & Co.) for some interesting views on technology and education.

"I believe in giving young people the freedom to chart their own learning paths," Claudia says. "The online world makes this possible now in a way that was not possible before. Online savviness is the new basic computer skill. Kids don't need to be taught online skills. Give them computers with modems, ISDN (high-speed, high-bandwidth data transmission), $99 Connectix video cameras, the CU-SeeMe software that's available free on the Internet, Netscape software, and get out of the way. Pretty soon, they'll be showing you things you never even dreamed of . . . . Once you've seen this combination in action, there's no going back. People will gather and collaborate around shared interests and ideas all across the planet."

# Highlights: Courses and degrees available online

Some of the following are partial excerpts from *The Electronic University,* copyright 1993 by Peterson's Guides, Inc. Princeton, NJ 08053-2123, and are used by permission. For more information, please visit their Web site (see Fig. 9-1).

## Agriculture

Colorado State University
SURGE Coordinator, Division of Continuing Education
Colorado State University
Fort Collins, CO 80523
Phone 303-491-5288, fax 303-491-7886

Master's or second baccalaureate in agricultural engineering

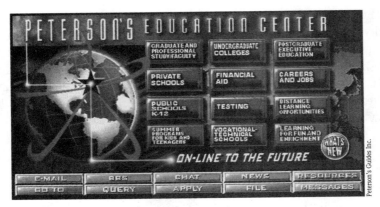

**9-1** *Peterson's Electronic University on the Web.*

## Behavioral sciences

New York Institute of Technology
On-Line Campus, Room 417
Old Westbury, NY 11568
Phone 800-222-NYIT or 516-686-7712, fax 516-484-8327

Formerly known as the American Open University. Private, fully accredited. Offers Bachelor of Science in Behavioral Sciences with majors in psychology, community mental health, sociology, and criminal justice.

## Business and nonprofit organization management

New York Institute of Technology
On-Line Campus
Old Westbury, NY 11568
Phone 516-686-7712 or 800-222-6948, fax 516-484-8327

Formerly known as the American Open University. Private, fully accredited. Offers Bachelor of Science in Business Administration with a management focus option and related business courses.

New York University
Informational Technologies Institute
48 Cooper Square
New York, NY 10003
Phone 212-998-7190, fax 212-995-4131
E-mail: vigilante@acfcluster.nyu.edu

Graduate-level courses in business management. Taught partly through compressed-video television.

State University of New York (SUNY)
Empire State College, Center for Distance Learning
Saratoga Springs, NY 12866
Phone 518-587-2100, fax 518-587-5404
Bitnet: CDL@SNYESCVA

Bachelor's degree in Business Management.

Thomas A. Edison State College
CALL (Computer-Assisted Lifelong Learning) Network
CALL Network Technical Center
101 West State Street
Trenton, NJ 08608-1176
Phone 609-777-4140, fax 609-633-6463
Internet: Telnet to CALL.TESC.EDU
For info by modem: 609-292-7200 (8N1, VT100, 2400 bps)

Fully accredited. For-credit courses in management and marketing, and will soon be offering a Master's in management.

University of Phoenix Online
100 Spear Street, Suite 200
San Francisco, CA 94105
Phone 415-541-0141 or 800-742-4742, fax 415-541-0761
On CompuServe: GO UOP

Private, fully accredited. Offers bachelor's degrees in Business Administration or Management, and MBAs, including Master of Business Administration, Master of Business Administration in Technology Management, and Master of Arts in Organizational Management. Students enrolled in courses can access their online computer network, Alex, either through CompuServe or directly.

# Computing and computer-related subjects

George Washington University
GW Television, Academic Center T306
Washington, DC 20052
Phone 202-994-8233, fax 202-994-4048
Bitnet e-mail: TOMWING@GWUVM

Master's in Electrical Engineering/Computer Science. Taught partly through cable television.

International School of Information Management
University Business Center, P.O. Box 1999
Santa Barbara, CA 93116-1999
Phone 805-685-1500 or 800-441-4746, fax 805-685-9685

Graduate-level courses designed for working adults with experience in information management. Offers Master of Science in Information Resources Management and a Master of Business Administration with Information Systems emphasis. Noncredit and custom-designed training courses for corporate staff are also available.

New York University
Informational Technologies Institute
48 Cooper Square
New York, NY 10003
Phone 212-998-7190, fax 212-995-4131
E-mail: vigilante@acfcluster.nyu.edu

Master's in Management/Computer Science. Taught partly through compressed-video television.

Nova University
Center for Computer and Information Services
3301 College Avenue
Ft. Lauderdale, FL 33314
Phone 305-475-7047 or 800-541-1682 ext. 7047, fax 305-476-1982
Internet e-mail: doctor@novavax.nova.edu

Private, fully accredited. Offers master's degrees in Computer Education, Information Technology and Resource Management, Information Systems, and Training and Learning, as well as doctorates in Information Systems, Information Sciences, Training and Learning, and Computer Education. Requires two one-week institutes at the Fort Lauderdale campus.

University of California, Santa Barbara
Student Affairs Officer, Off-Campus Studies
Santa Barbara, CA 93106
Phone 805-893-4056

Master's in Electrical and Computer Engineering, Master's in Computer Science.

# Education

Boise State University
Instructional and Performance Technology Department
1910 University Drive
Boise, ID 83725
Phone 208-385-1899, fax 208-385-4081

Master of Science in Instructional Performance Design and Technology.

George Washington University
GW Television, Academic Center T306
Washington, DC 20052
Phone 202-994-8233, fax 202-994-4048
Bitnet e-mail: TOMWING@GWUVM

Master's in Educational Technology. Taught partly through cable television.

University of Alaska, Fairbanks
Off-Campus Programs, School of Education
706 C Gruening Building
Fairbanks, AK 997755
Phone 097-474-6431, fax 907-474-5451

Baccalaureate in Education, master's in Education with specialities in cross-cultural education, curriculum and instruction, language and literacy, and educational leadership.

# Engineering

Rochester Institute of Technology
Information Technology
P.O. Box 9887
Rochester, NY 14623
Phone 800-CALL-RIT, fax 716-475-7100
E-mail: phl@cs.rit.edu

Master's in Software Engineering, baccalaureate in Electrical/Mechanical Engineering, master's in Computer Science.

University of Massachusetts at Amherst
Marketing Coordinator, Video Instructional Program
113 Marcus Hall
Amherst, MA 01003
Phone 413-545-0063, fax 413-545-1227
E-mail: Bowman@ECS.UMASS.EDU

Master's in Engineering, taught partly by TV, video, phone, and fax, as well as e-mail.

# Journalism

University of Memphis
See CompuServe, later in this chapter.

# Liberal arts

Atlantic Union College
Electronic Distance Learning, Box 1000
South Lancaster, MA 01561
Phone 508-368-2394, fax 508-368-2386

Associate Degree in General Studies.

University of California, Santa Barbara
Student Affairs Officer, Off-Campus Studies
Santa Barbara, CA 93106
Phone 805-893-4056

Bachelor's in liberal arts.

# Sociology, social work, and related subjects

Connect Ed (Connected Education, Inc.)
65 Shirley Lane
White Plains, NY 10607
Phone 914-428-8766, fax 914-428-8775
CompuServe: 72517,3107 and Internet: plevinson@cinti.com

Affiliated with The New School for Social Research, in New York City, although Connect Ed is a privately owned, independently run service. Credits for Connect Ed's Technology in Society count toward The New School's Master of Arts in Media Studies, and all courses for that degree can be taken entirely online. Those and other courses may also be taken for undergraduate credit or as noncredit courses instead. Connect Ed also offers noncredit courses in writing, foreign languages, and online research.

State University of New York (SUNY)
Empire State College, Center for Distance Learning
Saratoga Springs, NY 12866
Phone 518-587-2100, fax 518-587-5404
Bitnet: CDL@SNYESCVA

Bachelor's degree in Human Services.

University of Alaska, Fairbanks
Off-Campus Programs, School of Education
706 C Gruening Building
Fairbanks, AK 997755
Phone 097-474-6431, fax 907-474-5451

Bachelor's in social work.

# Other subjects

California Institute of Integral Studies
CIIS-School for Transformative Learning
765 Ashbury Street
San Francisco, CA 94117
Phone 415-753-6100 ext. 263, fax 415-753-1169

Interdisciplinary Doctoral Studies Program, about change in individuals, groups, and classes.

New York Institute of Technology
On-Line Campus
Old Westbury, NY 11568
Phone 516-686-7712 or 800-222-6948, fax 516-484-8327

Formerly known as the American Open University. Private, fully accredited. Offers Bachelor of Arts in Interdisciplinary Studies, Bachelor of Science in Interdisciplinary Studies, Bachelor of Professional Studies in Interdisciplinary Studies.

State University of New York (SUNY)
Empire State College, Center for Distance Learning
Saratoga Springs, NY 12866
Phone 518-587-2100, fax 518-587-5404
Bitnet: CDL@SNYESCVA

Bachelor's degree in Interdisciplinary Studies.

Thomas A. Edison State College
CALL (Computer-Assisted Lifelong Learning) Network
CALL Network Technical Center
101 West State Street
Trenton, NJ 08608-1176
Phone 609-777-4140, fax 609-633-6463
Internet: Telnet to CALL.TESC.EDU
For info by modem: 609-292-7200 (8N1, VT100, 2400 bps)

Fully accredited. For-credit courses in social psychology, computers in society, global environmental change, international economics, etc.

University of Alaska, Fairbanks
Off-Campus Programs, School of Education
706 C Gruening Building
Fairbanks, AK 997755
Phone 097-474-6431, fax 907-474-5451

Bachelor's in Rural Development.

# Commercial networks that offer online courses and tutoring

Several commercial nets offer homework help for students, university courses for undergraduate or graduate degrees, adult-education classes, and seminars. Most are in the fairly early stages of development, but the smart ones will increase their options steadily.

*America Online* There are several noncredit and for-credit courses offered through AOL's Interactive Education Center Services (keyword: IES), ranging from how to use AOL's Internet services to genealogy to university-sponsored courses. The Academic Assistance Center (see Fig. 9-2) has teachers available for homework help, including term-paper research guidance through their Academic Research Service. The Exam Prep Center helps students with study skills and preparation for standardized tests, such as the GED and SAT.

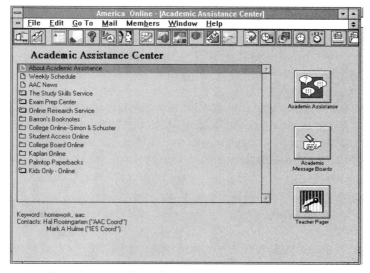

**9-2** *America Online's Academic Assistance Center.*

*CompuServe* The University of Memphis offers courses for a master's in Journalism through the Journalism Forum (GO JForum or send e-mail to Bill Brady, Ph.D., 70117,1460) and public relations courses through the PR and Marketing Forum (GO PRSIG). *Freelance success* newsletter sponsors a four-week course, "Magazine Writing Successful Freelance and others" (e-mail to 74774,1440 or 70421,2063) and the PR and Marketing and Working from Home forum both hold periodic seminars that stretch over several days or weeks, as well. Homework help is available to an extent, through the Students Forum (GO StuFo).

*GEnie* Several academic courses, including English, math, science, and social sciences, as well as adult-education classes and self-study courses to prepare for professional licensing exams, such as the real estate broker's license, are available through the Computer-Assisted Learning Center (CALC). Among the commercial services, GEnie was a pioneer in online education. CALC is accredited and has been operating since 1986. Page keyword: education or CALC.

*Prodigy* Prodigy offers additional software for their Homework Helper service, which searches selected reference publications to assist students working on papers or just studying for a class or exam. The software is free, but the service costs extra, although there is a free trial period.

*Women's Wire* Here you'll find occasional seminars or workshops on specific topics led by professionals in the field, such as personal finance, or sponsored by other companies or publishers. Wire just added this service in 1995, and they plan to increase the frequency and eventually offer classes.

# CHAPTER 10

# Pleasurable pursuits

**T**ravel, movies, music, hobbies, and sports are like sex. People love to talk about them almost as much as they like actually taking part in them. So it's no surprise that all online networks offer multiple forums for each of those topics, and that they're among the most active.

*Online it's like a club, and you know people have all come to talk about that specific topic.*

"If I see a movie I like a great deal, I want to talk about it with somebody," says Ellen Connally, a municipal court judge in Cleveland. "If I call up my friend, she may be busy with her kids. But online it's like a club, and you know people have all come to talk about that specific topic." Ellen often logs on to catch up on her e-mail at 5 A.M., before the intensity of her day demands every minute of her time. But contrary to the image most of us have of judges as austere serious people, she has also been known to siphon off the stress of her job by playing games online.

Because the people who frequent leisure-time forums like to do things rather than just sit around and type, participating even in the games forums is by no means a modem-to-modem experience, as Eliza Intino's story shows. When Eliza, who lives in L.A., ventured into one of the many game areas that are ubiquitous online, she found it pretty confusing at first. Leonardo Drago went out of his way to help her, time after time, and she grew fond of him even though they'd never met or talked by phone. By then, she knew that he was a college student from Hong Kong, but hadn't realized that he was studying in the U.S., at Boston University. When she heard that, it made her think about how alone he must feel sometimes, in a foreign land so far from home.

"Leo brought out the mother instinct in me," she says, "My husband and I never had children, so I jokingly adopted him one day." Leo and other members of the forum dubbed Eliza "Mom USA." Just like any doting mother with a son away at college, she began sending him holiday and birthday gifts and frequent care packages crammed with cookies or homemade bread, and they eventually communicated daily in chat mode. "My husband kind of adopted him too," she says. "He won't go near the computer, but he always says, 'How's Leo?' or 'Tell Leo I said hi.'"

To find games and game forums on any commercial service or the Internet, just use *games* as a keyword.

# Travel tales

Members of the travel forums, especially, frequently visit one another during their trips. They also share in each others' journeys by giving recommendations for what to see, where to stay, and which restaurants to try, then reporting back to everyone when they return. So it becomes a shared adventure even for those who can't travel often themselves.

Paula Beasley, of San Juan, Puerto Rico, says she hangs out in the travel and foreign-language forums as a way to expand her world. "When I was younger I travelled some, especially in Europe. I have always had an interest in Europe and the U.K. and really enjoyed my trips. Now it isn't possible for me to pack a bag, take a cheap flight to some other country, and ride trains from country to country. Instead, I joined the European Forum, the Italian Forum, and the U.K. Forum. I talk to people who live in other countries or who have visited there. It isn't quite the same as being there, but it is interesting and brings back great memories. I really enjoy it, and virtual travel beats no travel at all!"

– 151 –

The information available both in the forum libraries and from other members have saved many a trip from impending disappointment or cancellation, as Thomas Doughty attests. He and his sister had agreed to drive their mother and aunt around the Alsace region of France. Because his mother and aunt live in Wisconsin and he lives in California, they had all decided to meet in Frankfurt and proceed from there. At the car rental counter in Frankfurt, he discovered that his mother and aunt thought he had mapped the route and he thought they had, and the result was that they didn't know where they were headed. Fortunately, Paul had logged onto the Travel Forum on CompuServe the night before he left and downloaded a diary by Paul Muns, which described his week-long bicycle trip through Alsace with several friends.

"I led my little troop from Mutzig to Obernai, Ottrott to Schirmeck," says Thomas. "Following Paul's bike route, we stopped for a meal at Château de

Barembach," which General Patton used as the third Army headquarters during the final days of World War II. Paul's log had included the name of the proprietress, so Thomas and his entourage introduced themselves and explained how they had found her. She remembered the cyclists because 24 people taking showers at once had overloaded the château's water pipes. But she laughed about it, and invited them all back.

"She then augmented Paul's diary into a fairy-tale itinerary for us: more wonderful villages and trails, with splendid overlooks at every turn," Thomas says. "Between Paul's diary and Madame Clement's map, my mom and aunt shared breathtaking views of upland meadows, forested hills, and terraced vineyards, all dotted with storybook castles and monasteries. It was a banner trip."

There are many stories about travel forum people helping each other out when they encounter roadblocks, even when they know little more about each other than their online addresses—and, in Scott Hatton's case, when they don't even know that. Scott, who lives in Dundee, Scotland, got a last-minute invitation to spend Christmas in Morocco and really wanted to go, but all flights were booked. His only hope of getting there in time was a flight to southern Spain on Christmas Eve, then a ferry from Gibraltar to Morocco, but he had no idea how to check ferry schedules or whether they even ran on Christmas Eve.

"I had a brainwave," he says. "I figured Gibraltar was bound to be full of people called Smith and Jones. I simply did a search for *Smith* and *Gibraltar* in the main Members Directory on CompuServe, and found Smith, Louis T., of Gibraltar. Feeling I was taking a bit of a liberty, I e-mailed him."

Louis T. Smith went to considerable trouble to find a ferry that would get Scott to his destination, even though none were available near the airport where he was to land, and e-mailed the timetable back to him. "On Christmas Eve night I was supping Moroccan beer in the Rif Hotel in Tangier," says Scott. "I had a wild time in Morocco, instead of a normal Christmas in the frosty, foggy U.K. All thanks to Louis."

To find travel forums on any commercial service or the Internet, just use *travel* as a keyword. Other logical words will also lead you to related sources, such as *parks, outdoor, hotel, cruise, resort,* and *recreation,* particularly on the Web. On the Internet, also look for the rec.travel newsgroups.

# Fabric arts and other crafts

Members of crafts forums help each other on projects regularly, and frequently work together on group projects such as the progressive quilt shown in Fig. 10-1, made by members of CompuServe's Fibercrafts Forum, who communicated by e-mail and earth mail. The four women who made this particular quilt live in different parts of the U.S., which is why they call it the Continental Drift quilt. Each created a different design, but each woman also made four blocks so they could all have complete and identical quilts. Others made a baby quilt as a surprise gift for a member's baby shower, and officially presented it to her online (then offline, of course), and uploaded photos of it so all the other members could see it and be part of the event too.

**10-1** *The Continental Drift progressive quilt, made by four members of the Fibercrafts Forum on CompuServe.*

Any self-respecting quilt lover knows that the thin, machine-made kind you buy at department stores don't even come close to the thick, soft quilts that have love hand-stitched into every square. Computer-drawn patterns for quilts have brought an age-old art into the technological era, and trading tips online has helped more people learn the art of quilting and helped keep the knowledge of a treasured folkcraft from dying out.

*Tip:* To find crafts forums online, simply use crafts or the name of any specific craft, such as quilting or pottery, as keywords.

# Genealogy

Many adopted children have found their birth mothers through quests that began online. Many people also use the member directories or numerous genealogy forums to search for long-lost friends, classmates, or relatives. On the Web, there's even an affordable fee-based service called Find-a-Friend. Sometimes, though, the connections happen by serendipity.

Mike Nellis, of Boston, says, "I was browsing the member directory and noticed my name listed twice. My last name is not that common, and I certainly didn't expect to see someone with the same first name, middle initial, and last name." My full name is Michael Patrick Nellis, and it turns out that there's another member named Michael Paul Nellis." Through e-mail with his namesake, Michael learned that "His father's name was David and his brother's name was Patrick. I have a brother named Patrick, and my brother's middle name is David. My confirmed name is also David. I go by Mike online, and he goes by Michael, but all those Irish names are quite a coincidence."

Their names weren't the only coincidence, says Mike. "He is in the Navy, working in the electronics field and stationed in Adak, Alaska. I was in the Navy for four years and worked in the electronics field. We're two years apart in age. His father and my father were both born about the same time in the Rochester and Syracuse, New York area, but apparently never knew each other. We lost touch when he was transferred to a ship, but it was a real case of eerie dé ja vu."

Les Rix, of the U.K., also discovered someone with the same name, and a surprising connection. Roger Rix, of Eugene, Oregon, found Les's name in the directory and wrote him a note. They've since discovered that they not only have a mutual interest in CB radio, but, sure enough, they might have mutual ancestors. Through their online correspondence and Les's diligent search of records in England, they've learned that Roger's great grandfather emigrated to

Warsaw, New York in about 1882 from Watton, in Norfolk, a small English village close to where Les was born.

"East Anglicans, even today, are reluctant to uproot," says Les. "My own forebears never left the county. I feel certain that his granddad would have left England due to the terrible slump in agriculture in the late 1870s. Whilst some village groups went en masse to Canada or Australia, we are very interested in learning what Charles did for a living, and why he decided to settle in the U.S."

Les has since sent copies of historical records and photos to Roger, and they've continued their mutual ancestor search through both e-mail and CB radio. They think they're fourth-generation cousins. "I am delighted at meeting Roger this way," he says. "My son is suddenly aware that the U.S. is something more than just a place on the news."

Because it's so natural to use computers for research, there are genealogy forums on most services. America Online offers genealogy classes through their Interactive Education Service. McGraw-Hill also has a book on online genealogy resources, by Elizabeth Crowe, called *Genealogy Online: Researching Your Roots.*

# The arts

Among Internet newsgroups, forums on commercial services, and sites on the World Wide Web, you'll find discussions and abundant information—even sound and video clips—for just about any kind of music. There are also listings of concerts and festivals worldwide (the Yahoo search engine one on the Web is particularly good), and listings by city or region for bands' club performance dates and plays. CompuServe and eWorld have good forums for musicians as well as fans. Every network also has sections for writers and avid readers, so the literary, visual, and performing arts are all well-represented.

Visual artists are making especially good use of the medium by displaying their art electronically through online artists' cooperatives and Web sites sponsored by private and university galleries. Since this is unjuried self-promotion in some cases, the quality varies greatly, but it's a way for new artists to get discovered and for you to discover artists while their work is still affordable. Figure 10-2 is the opening screen from Muriel Magenta's World Wide Web artists' cooperative, The World's Women On-line! It features various kinds of art from women throughout the world, which she plans to display by video projection at the United Nations' Fourth International Conference on Women, in Bejing in September 1995.

**10-2** *The opening screen from Muriel Magenta's World Wide Web artists' cooperative, called The World's Women On-Line!*

However, it would be hard to top the resourceful way that John V. Scialli, a psychiatrist in Phoenix, dreamed up to use the online services. He's an avid fan of Frank Zappa, the late rock musician, and felt that Zappa was such a shining star of 20th-century music that he deserved to have a real star named in his honor, as a memorial. So he waged a campaign in 15 countries to get Dr. Brian Marsden, director of the International Astronomical Union's Minor Planet Center, to name an asteroid discovered by the Czechs after Zappa.

"I networked electronically, by using e-mail and posting messages in the RockNet and Music/Arts forums on CompuServe and on alt.fan.frank-zappa, a Usenet newsgroup on the Internet, John explains. He even got a letter of support from the Czech government. "Dr. Marsden received hundreds of endorsements by e-mail and fax." He said it was the largest lobbying effort he had seen in the naming of more than 3,000 minor planets."

John is proud to report that a five-mile-across asteroid orbiting between Mars and Jupiter is now named Zappafrank.

*Tip:* Keywords that work, among others, are: *music, art, arts, TV, television, theater, theatre, writing, writer, writers, literature, museum, gallery, dance, Hollywood, entertainment, film,* and *movies* (with searchable, detailed databases and reviews to keep any film buff occupied for hours). Most commercial services have a choice on the main directory called Entertainment or Leisure. Also check the Expo and WebMuseum sites on the Web.

# Participant and spectator sports

Monday-morning quarterbacking is as prevalent online as it is elsewhere, but here you can also be part owner of an entire league, and negotiate the players' strikes, trades, and deals with other members. It's all a simulated game of course, but it's also all from the comfort of your keyboard, so you're not risking millions nor alienating your minions. And it's a heckuva lot more fun when you have knowledgeable people on the other side of the bargaining screen, ready and willing to resume the talks whenever you are. Anyone who thinks only men are truly interested in major-league contact sports hasn't been paying attention to trends of late.

But if tennis or backgammon are more your style, you can find tennis partners online, learn what to do about ankle or elbow injuries, and play a game of backgammon, playing either both sides of the board or with a partner. If you're more the hiking or biking type, you'll find many others who share your proclivity and who can tell you where the best trails are practically anywhere in the world.

There's something for everyone, but it's not always in the most obvious place. Check the rec. hierarchy of newsgroups on the Net and use *sports, games, health, fitness, travel,* or the name of any specific sport as keywords to explore what's available on the Internet and the Web, as well as the commercial services. *Tip:* On the Web, try the World-Wide Web of Sports at: http://tns-www.lcs.mit.edu/cgi-bin/sports.

# The World Wide Web as entertainment in itself

More lethargic types, like myself, are perfectly content with less strenuous activities or adventures of the mind. So if history, photography, or even extra-terrestrial intelligence research is your avocation, you'll find a niche online (to make it easy, the last is on the Web at http://metrolink.com/seti/SETI.html). Some of us also find it fun to occasionally see what we can discover without looking for anything specific. The best place to do that is on the World Wide Web and there are two great ways. The first is the Random Links menu option in Yahoo. (Just type yahoo into your Web browser and then save it with the hotlist or bookmark feature, because it's an all-around great resource, even though there's now a nominal extra charge.) The second safe bet is by playing Web Roulette (see Fig. 10-3) at: http://kuhttp.cc.ukans.edu/cwis/organizations/kucia/uroulette/uroulette.html

If you prefer real roulette, there will be a casino opening on the Web in the summer of 1995. Because it wasn't available when this book went to press, it remains to be seen whether they'll also use real money. Some people believe it is a gamble to use any kind of money online (which should change by laet '95 say others).

About 4,000 individuals and families so far have turned the Web into recreation in itself in another way, with personal, or vanity, Web pages. Some are boring and a total waste of time; some are silly, but clever; and some are useful to others, such as the Farmer family's home page at http://www.infi.net/~dolores/, created by Dolores Farmer. It's a combination tour guide to the Roanoke, Virginia area and an introduction to her family and the way they live.

University of Kansas

**10-3** *Web Roulette. Just click one of the arrows, and see what comes up.*

"Try things. Take a chance!" advises Dolores. "I have no computer training whatsoever. Not one class. I wrote the page myself, but we consider it a family project. I do all the HTML writing [hypertext markup language, or text conversion and links to other pages], choose the graphics, and so forth. My husband Eddie often contributes ideas about subject matter. Our son, Justin, is only 2, but he already works a mean mouse. He's also the reason we have a link to 'Blue, the counting dog.' Blue barks out the answer to simple math problems. Kids just eat it up.

"I wrote the page because I am simply stunned at the reach of this technology. It is still hard for me to believe the astounding fact that I, in effect, have become an international publisher. That used to be the province of McGraw-Hill, not Dolores Farmer! Last week, a man named Pascal LaMeur filled in the form on my Web page. I am in Salem, Virginia. He is in Lausanne, Switzerland. This week, a man from Germany sent e-mail saying he enjoyed the Farmer Family page."

*I wrote the page because I am simply stunned at the reach of this technology. . . . I have been touched by other people's home pages, and they by mine.*

"I saw a quote somewhere on the Web that said: 'Teach someone to surf the Web and they can *tour* the world; teach someone to publish on the Web, and they can *touch* the world,'" says Dolores. "Corny, perhaps, but true. I have been touched by other people's home pages, and they by mine. Certainly, I never received letters from people all over the world before I published a page.

"Since I published a page, I have been asked by both my employer and my alma mater to write their pages. I have met people that I never would have met otherwise. A number of employment opportunities have arisen that I never could have imagined. And all that doesn't even take into account the page appearing in this book. Every time I connect to the Internet, it's a new adventure. I never know what's waiting for me."

If you stop by the Farmer Family Home Page, tell Blue I sent ya'. There's a guest book for visitors to sign, so please say hello to Dolores too.

# CHAPTER 11

# Psyche and spirit

So many messages online are supportive, and it's such a medium of ideas and mind-to-mind communication that, on the surface, it would also seem a natural place for the communion of souls. And it is. But like much else online, you'll seldom find it in the obvious places or in the obvious ways.

## Seek and ye shall find

"I've had some wonderfully enlightening experiences online," says Mike Bayer, a minister who also happens to be a sysop for an unrelated forum on CompuServe, "however, they all took place behind the open message areas. I've prayed online (typing my prayer onto the screen), and been prayed for. Usually, what happens with me is I connect with someone online, and then exchange a message in private."

The wariness people usually feel about baring their souls to strangers online is warranted, just as it is in the real world, Mike says. "This is an area that can be abused easily. People tend to be either defensive or vulnerable in these forums, and the vulnerable ones have to be careful whom they share with. The positive thing going on in cyberspace for religion is the availability for people of like faith and conviction to hook up on a worldwide basis. If you're looking for Baptists, Orthodox Jews, or New Age gurus, you can find someplace where they gather. Whether you fit in with them is another story.

"The online forums for religion suffer from the same pains that organized religion suffers from—having to contrive something when it appears nothing is going on," Mike continues. "But there are little pockets of spirituality hidden in other forums online. One of them is the poetry section of the Literary Forum on CompuServe. Without an expressed purpose of being spiritual, many of the poetic posts are very spiritual and healing in nature. Some of the richest recovery and spiritual poetry I've read has come out of this area.

"People of faith, religion, and spirituality are in every forum, and some of the best spiritual experiences I've had have been in the professional and hobby forums. If you buy what people like Marianne Williamson say, then there's the potential for spiritual energy in any connection that we make. I would rather find a soul-seeker in the Journalism Forum who shares some of my professional and spiritual interests than find a same-kind Christian in a religion forum who disagrees with everything in my life except my faith."

Despite obvious attempts to make them closer to all-inclusive, the sections in religion forums on all the commercial services lean toward Christianity, with America Online having the strongest Christian bent, largely because *Christianity Today* and *Christian Woman* have their own sections there.

There are multiple places for every conceivable faith, denomination, and spiritual practice, however. Information and opinions are abundant, and you'll have access to perspectives and people beyond what most of us would find in one locale without great difficulty. But much of the information, particularly on the Internet, is rather academic, and the opinions everywhere are too often vociferous and rigid. Within women's and New Age-type forums, the slant is often toward wicca (Goddess worship) and witches. Elsewhere, you get the sense that the forums and newsgroups about religion are bastions for those who are comfortable in their beliefs and chosen religion, rather than those of us who feel we're still seeking rather than feeling sure we've found the way, the truth, and the light.

# Two typical discussions about spirituality

Two discussions on Prodigy that happened to be going on in different forums as I was writing this chapter (March '95) are pretty typical of what you'll find in the realm of spiritual discussions. Religion Concourse 2 on Prodigy tends to be about alternative spiritual practices, rather than the more mainstream or recognized religions favored in Religion Concourse 1. I haven't the vaguest idea what concourses—which remind me of airports or shopping malls—have to do with religion, although that title fits with the messages about travel deals and investment bonanzas that keep popping up at the bottom of the screen, as they do all the time on Prodigy, everywhere.

The garish ads are even more disconcerting than usual in the context of religion, however.

The consensus in the discussion in Religion Concourse 2 on whether it's possible to find online spiritual guides or mentors, and how or whether online networks can be useful to spiritual seekers at all was both yes and no. You can find some kindred spirits who can help you or whom you can help on their journeys and, like most other things of real consequence online, that relationship might begin because of a public discussion. But it will most likely develop from there in private, through e-mail, as Mike Bayer says.

There are exceptions, or attempts to reach out and risk. When the Spirituality section of the Women's Leadership Connection on Prodigy began, for instance, nobody talked about how enlightened they were, nor how humble, nor how they'd just discovered a great guru. Nobody mentioned a single church fundraising campaign nor a word about their interpretation of some scripture. Instead, one member told of the unsettling coincidences that had happened to her over the past few months that had filled her with an eerie, disturbing premonition that she was going to die soon, before her 45th birthday. Another talked of a neighbor who had recently died, and how that brought up all kinds of questions about an afterlife and whether any of our lives really mean anything, and how difficult it was for her to reassure her son when he asked if his friend was now in heaven. She had no idea, really, she lamented, and was filled with doubts. She asked if any of the rest of us ever had those kinds of doubts when a loved one died, and whether we'd found any answers or a plateau of peace.

The first few women who responded to that message replied heart-to-heart, soul-to-soul. It felt like they were willing to expose their innermost thoughts to strangers partly because they were strangers, and partly because this forum somehow felt more sheltered and private and less pretentious than the more public ones, where people don't have to apply for access as they do in the Women's Leadership Connection. Alas, other people felt uncomfortable with the personal nature of that exchange, and quickly managed to change it into a discussion of the comfort pets can offer and then, of all things, looking for dogs to buy or adopt.

# Where support groups meet 24 hours a day

Just as dark nights and light come to different souls in different ways, forums designated as places to discuss spiritual matters aren't the only ones online where people seek sustenance for the psyche, says Mike Bayer. Communication by modem is a near-perfect medium for unburdening yourself and finding kindred spirits. They're online at 4 A.M. when it feels like you'll never make it through the night, but feel you can't call anybody. The bonds that develop between people connected by modem can be every bit as strong as those among members of any group that meets in someone's living room or a church basement.

Many of the online support groups are out of public view or awareness, however, because they're either private Internet mailing lists or small groups that have formed and meet in a private chat room or an Internet relay channel at a designated time every week or even every night. So it would be tough for anyone to come up with a truly good, much less definitive list. Just the same, what follows is a useful starter kit.

– 165 –

# Highlights: Online spirituality discussions and support groups

The best places for substantive discussions of personal concerns are in closed sections, whether they be Internet mailing lists, women-only sections, or health forums on any of the networks, or invitation-only chat sessions.

## Recommended

*The Transformations forum on eWorld* This is the only forum on the commercial networks that was created specifically as a self-help forum. It's run by Becky Boone, a psychologist in private practice who also started the Women Online Worldwide forum on eWorld (see chapter 20 for more information on that forum). There are sections for 12-step programs, medical problems, mental health self-help, and several support groups that meet regularly, as well as the daily discussions in those sections. (See Fig. 11-1.)

**11-1** *The Transformations Forum on eWorld.*

**The Better Health and Medical Forum on America Online** Of the major commercial networks, America online gets points for support groups that hold weekly sessions in the Issues in Mental Health and Better Health and Medical forums, but not so much its more obvious sections, and certainly not for the typical People Connection chat rooms. The Better Health and Medical Forum hosts regularly scheduled self-help, real-time chats for people coping with attention-deficit disorder, obsessive-compulsive disorder, and autism, and there's also an Abuse Survivors' Self-Help group that meets weekly, as well as many other support groups that are physical-health oriented. In the Issues in Mental Health Forum, you'll find regular message board sections on daily living, divorce and separation, and depression, and daily or weekly support groups titled Anxieties and Phobias, Stress Busters, Sober Today, ADD Support (separate groups for adults or parents of children affected by attention-deficit disorder), Widowed World, Starting Over, Work Day Woes, Counselors' Corner, Eating Recovery, One Day at a Time, Divided Minds, Anxiety and Phobias, Depression Support, Mood Disorders, Panic Disorders, and others.

If you wend your way to the Support Groups section, first via AOL's Exchange then, through the Communities Center, you'll find more than 40 discussions on different topics, ranging from suicide support to divorce, all started by members looking for others with empathy. There's also a Religion and Ethics Forum, and John Battle teaches an eight-week course on the Life and Letters of St. Paul at the Electronic University of the Education Forum, for only $25. "This course is especially helpful for Sunday school teachers and Bible study leaders, and for anyone seeking to understand the Bible better," he says in his introduction.

*Internet mailing lists* Judy Heim, who had a brain tumor that irrevocably changed her life, says she relies on the Brain Tumor Support List as her online support group. "The thing I like about it is that other people have gone through the same things I have—losing their jobs or not being able to return to work full-time because of a tumor, and all the financial insecurity this leads to. It's been a tremendous comfort knowing that there are others going through the same thing. There are some pretty heavy philosophical and spiritual discussions going on too, because so many members are facing death or have children who are dying." For information on how to find mailing lists on the Internet, see appendix G.

# Worth exploring

*CompuServe* Here, the Health and Fitness Forum is a key place because it has a self-help orientation, with weekly AA meetings and a section run by Ed Madara, Director of the American Self-Help Clearinghouse. He's also the editor of the Clearinghouse's *Self-Help Sourcebook*, and says the next edition will give guidelines on how to start an online support group. See the Bibliography for information on the book. The Human Sexuality and Relationships Forums discussed in chapter 6, *Sex and romance*, are definitely worth investigating, too.

*Delphi* Read the roster and you'll find several good areas for the spirit and psyche among the custom forums, all created by members. Among them are the Single Parents Network; Explorations of Life; the Codependency Support Group; Divorce Support Board (before, during, and after, it says); a 12-step program for sex or love addiction; and the Depression Support Group. There are also the National Philosophy Forum, the Metaphysical Universe, and a few religion forums running the gamut from Christian to pagan, with a pretty active one for Unitarians.

*GEnie* The Family & Personal Growth Bulletin Board hosts a closed section on abuse called Survival and Recovery, where admission is granted on request after you agree to policies intended to protect people's dignity and create a supportive atmosphere. There's also a closed section titled Women Only that could be good for getting and giving feedback. Of course, there are continual discussions on the public board in various self-help topics, as well.

*Prodigy* As mentioned earlier, Prodigy's Religion Concourse 1 has sections for mainstream and not-so-mainstream but recognized religions or denominations, whereas Religion Concourse 2 provides a connection for less mainstream denominations, such as Unitarian Universalists, or other outlooks, such as New Age philosophies and practices. The Women's Leadership Connection, launched in late 1994, has a Spirituality section, and the Homelife Bulletin Board has a religion section. Among the topics on the Health Bulletin Board are emotional therapy, sleep and dreams, and stress management.

*The Internet and World Wide Web* In addition to the mailing lists mentioned earlier, there are several read-only gopher resources worth searching for. One that could become one of the better ones, although it just began in early 1995, is GriefNet, which encompasses death, dying, bereavement, and loss resources. The gopher address is gopher.rivendell.org.

The other place to look for support groups on the Internet is any of the newsgroups that start with alt.support. Among those, all of which are pretty self-explanatory, are alt.psychology.help, alt.support, (for emotional support in general), alt.support.attn-deficit, alt.support.big-folks, (for those who feel discriminated against because of their size), alt.support.depression, alt.support.diet (described as "seeking enlightenment through weight loss"), alt.support.divorce, alt.support.eating-disord, alt.support.shyness, alt.support.stop-smoking, alt.support.stuttering, and alt.support.tall. If you can't find the kind of group you're looking for, learn your way around and get a feel for how the Internet newsgroups operate, then propose one. Anyone can. If enough people are interested, it will happen. But be sure you search thoroughly first to make what you want doesn't already exist, because new groups start frequently. To find out how to join them, check with your online service provider.

Surprisingly, there's still not much on the Web, but that's because support groups require interaction and people are just beginning (as this book goes to press) to design truly interactive Web sites where people can communicate with each other rather than just look at things or be passive tourists. There are good omens, though. You can find Alanon and Alateen information at http://solar.rtd.utk.edu/~al-anon/. And, although it was pretty sparse when I checked, the Guide to Emotional Support on the Internet could develop into a useful reference point. Its address is: http://asa.ugl.lib.umich.edu/chdocs/support/emotion.html.

Religion on the Web is a good place to start to find out what's where. You can find it at: http://www.einet.net/galaxy/arts-and-humanities/religion.html.

For another wide-ranging directory of sites related to religion, begin at the Yahoo Religion page: http://akebono.stanford.edu/yahoo/society_and_culture/religion.

There's always something for people of all kinds online, so if the more mainstream faiths and philosophies aren't for you, check out Spirit-WWW, which has sections on UFOs, astrology, out-of-body experiences, yoga, channelings, and spiritual healing. The Web address is: http://err.ethz.ch/~kiwi/Spirit.html.

The Self-Help Center is an excellent model for what could be done nationwide or regionally to help people find support groups both online and offline, but this one's useful only to those in east central Illinois, so far. If that's you, you'll find it on the Web at: http://www.prairienet.org/community/health/self-help.html.

For those interested more in abstractions, download Philosophy in Cyberspace: A Guide to Philosophy-Related Resources on the Internet, a directory compiled by Dey Alexander of the Philosophy Department at Monash University in Melbourne, Australia, and available from The University of Michigan's Subject-Oriented Clearinghouse (see appendix G). This comprehensive list covers newsgroups, mailing lists, and more. For a whole shopping list of places to investigate, go to McGill University's Web site called Philosophy on the Internet. To get there, use your Web browser to type: http://godel.philo.mcgill.ca/philosophy.html.

The University of Chicago Philosophy Project on the Web is another good starting point: http://csmaclab-www.uchicago.edu/maclab/philosophyProject/.

**The WELL** This text-based BBS deserves inclusion here because WELLbeings love nothing better than a good debate or philosophizing, and folks in the San Francisco Bay area, the WELL's home, also place a high priority on things spiritual. In the Body, Mind, and Health section of the WELL, you'll find conferences called Buddhist, Christian, and Fringes of Reason, as well as ones on more general topics including mind, philosophy, psychology, religion, spirituality, and therapy. You might also want to check out the Life Stories conference in the Interactions section. But don't forget WOW, Women on the Well, and FemX, the place for GenX women. Problem-solving and emotional support are a major part of why they both exist.

# Honorable mention

**Women's Wire** The Women's Spirituality section on Women's Wire is in the Cultures and Communities section. There's not much going on here, but that could change as the network grows, and it's likely to grow a lot over the next couple of years. This is also where you'll find Charlitas, a special section for Latinas with all messages in Spanish.

# CHAPTER 12

# 100 ways to save time and money online

**Y**ou still haven't had time to finish reading last Sunday's paper, even though that's supposed to be one of your few leisurely indulgences. You got five more pieces of junk mail in today's delivery alone. You brought a briefcase full of reports home from the office to read before tomorrow morning's breakfast meeting. And the novel you ordered from the book club is still in the shipping package. More information clearly isn't something you need, so if you're skeptical about why you need to be online, no wonder.

Being online can add to the information overload, no doubt about it. You can fritter away hours just wandering the Web, and be utterly overwhelmed by e-mail if you're a prolific message writer or join too many newsgroups, forums or mailing lists. You might as well resign yourself to spending a big chunk of time the first few weeks you're online because it's like discovering a whole new world, and we all get caught up in the fascination. But rest assured that the novelty wears off. You learn to hit the Delete key without hesitation and develop other ways to control the influx of information, yet get the most benefit for the least time and money.

It's easy to get seduced by all the possibilities and just do more, more, more. But if that was the net effect of the Net, there wouldn't be 25 million people online. There are so many ways to save time and money, or just to do things more easily, that you'll soon find some of the traditional, earthbound ways a hassle and a waste of precious time. When it makes more sense to pick up the phone to call someone or see someone or someplace in person, do, of course. Computer-mediated communication, as the theorists call it, will never replace the value of direct contact with other people or seeing places first-hand. Used judiciously, however, it can give you more time to devote to the people and activities you really care about.

It isn't difficult to learn how to make the most of your time online, but it will take time initially just to explore enough to find out what's most useful to you. If you don't invest that time, you'll never quite see the pay-off. Books that give you lists of lists or tell you exactly how to find something online give you a quick fix, but that's not as helpful in the long run as learning how to find things yourself. So for each time- or money-saving tip that follows I've told you how to go about finding that information, but I intentionally don't include many specifics, although you'll find some of them in other chapters. That's partly because there are several ways to find the information or accomplish the task, and partly because the point is for you to learn how to find it on your own. It's a little like a scavenger hunt. Once you get the process down, a good way to learn more and have fun doing it is to try the mother of all online scavenger hunts, called The Internet Hunt. It's a weekly game in which you're given a list of facts or sites on the Net to find, and the first person to come back with everything on the list wins. Here's a clue for finding it: FTP to ftp.cic.net and get the file called INTRO.TXT. Happy hunting.

*1.* Book a hotel reservation or an ocean cruise. You won't be able to actually book a hotel or cruise ship reservation on the Net until late 1995, when they've finished installing security features, but that's in the works. Meanwhile, you can browse the equivalent of brochures with color photos, maps, and other details, then call the toll-free number to make reservations. *Hint*: Using either TravelWeb or TravelNet as keywords will get you to the right place on the Web (see Fig. 12-1). Also see chapter 10, *Pleasurable pursuits.*

**12-1** *Information on Le Mirador hotel in Geneva, Switzerland, available through Travel Web (http://www.travelweb.com).*

*2.* Schedule a flight. The Official Airline Guide and Easy Saabre—the same reservation system that a lot of the airlines and travel agencies use—are both available on some of the commercial services, and probably will be on the Internet eventually. Some services charge for access to them and some don't, so be sure to pay attention to the cost. They're a good alternative to travel agencies for after-hours planning or if you want to be more in control of your trip.

*3.* Check government travel advisories. If you're headed to a part of the world known for terrorism or civil wars, or one that recently had an epidemic, do not pass GO until you've checked the U.S. Government Travel Advisories update. They're available on several of the commercial services as well as the Internet.

*4.* Ask the locals for travel tips. No guide book can tell you what's really worth visiting in a city as well as the people who live there can. You'll find them in the travel forums on every major commercial service or in the Internet rec.travel newsgroups, as well as in several Internet newsgroups and commercial forums that focus on specific countries or regions. Civic pride makes them glad to brag about local attractions to anybody who stops by. Also see chapter 10, *Pleasurable pursuits.*

*5.* Go to a shopping mall without leaving home. Each of the commercial services has an online shopping area, and the Web has several, with more likely to come. They're still too luxury-, computer-, and text-oriented, but that will change as merchants get more sophisticated and confident about this new way of reaching buyers and as the technology evolves. CompuServe and Prodigy offer the most choices of the commercial nets, but America Online is expanding. All three offer or are soon likely to offer CD-ROM products that connect with their online shopping areas too, and provide extras such as video and sound. The Internet Mall, Commerce Web, Downtown Anywhere, and the Internet Shopping Network (online counterpart of the Home Shopping Network) are among the many shopping centers on the Web. Also see chapter 16, *Managing Your Money by Modem.*

*6.* Find phone numbers. Regional phone companies are now beginning to put their yellow-page listings on the Net, and it's only a matter of time until the white pages follow, although there will be a surcharge for those. You can already use Phone*File's white pages on CompuServe or their Biz*File business listings. If it worked as well in practice as in theory, the average cost of 25 cents per phone number found would beat the 75 cents to a dollar you end up paying by phone, but not once in the several times I tried it did I find the number for the company I wanted. It did work for finding residence numbers for two different individuals, however.

*7.* Have an in-home encyclopedia that's updated monthly. Now you can access the equivalent of a full set of encyclopedias whenever you need it, for a pittance, plus it's updated monthly. Several commercial services and the Internet have either Compton's Encyclopedia or Grolier's Academic American Encyclopedia. All you have to do is type in what you're looking for, and up it pops on your computer screen.

*8.* Do research for a report. Whatever you need to find, no matter how esoteric, you can probably find it or at least some reference to where to find it online. The trade-off is whether you want to spend time or money, because, until you get familiar enough with what's where and how to find what you need, you'll either pay surcharges to search databases on commercial services or roam the Net hoping to find what you want. For more information, see chapter 18, *Online research basics.*

*9.* Get tech support for computer hardware or software. Whether you want to master a desktop publishing program, find how to get your fax software working properly, or decide which laser printer or tax software program to buy, you can find help online, probably directly from the vendor's technical support staff. You might have to wait a day or three for an answer, but that's increasingly true of many of the phone hotlines as well. Be wary of taking advice from the many self-proclaimed experts online. You'll get so many different and contradictory answers that you'll only get more confused, and some of them won't be appropriate for the configuration of your system. For recommendations on what equipment or software others like, however, forum members can be very helpful because a consensus invariably emerges. The best service for help directly from the company has always been CompuServe, although Interchange and the Microsoft Network are likely to become key connections too.

*10.* Read newspapers and magazines. No publisher is foolish enough to think you're going to cancel all your subscriptions and read everything online, even if you have a laptop computer that you can take to bed or curl up on the couch with as you can a magazine or the Sunday paper. But none of us can afford the tab, time, or clutter to subscribe to everything we might occasionally find of interest, and we all occasionally want to read a story a friend mentioned that's in a magazine or newspaper we don't regularly read. That's where browsing online tables of contents and articles helps save both time and trees. Most of them have been free, but some are beginning to change to a subscription basis, and it's too soon to tell whether that will fly long-term or they'll have to try another tactic. When I subscribed to *Atlantic Monthly*, for instance, I never got around to reading much of it. Now I can check it on America Online to see if there's a story I want to read and print it out. For the shorter articles that most magazines run, I just scan or read the entire story online, then zap it. Gone! No clutter. Most major dailies are online, and many local papers already have online editions, some of which provide customized versions (see Fig. 12-2). Publications launch online editions every month. (Also see appendix F.)

**12-2** *The* Washington Post*'s Digital Ink, on AT&T's Interchange Online Network.*

*11.* Get customized news reports. There are several ways to get only the news you want culled from the major newswires, sometimes daily newspapers, and specialized newsletters, all delivered right to your computer at your convenience. While working on this book, for instance, I used HeadsUp, on eWorld, to follow just the news about the online industry and electronic publishing. Every morning, I got a two- or three-line summary of what was happening, and could request the full article for a nominal fee, or track it down through other, free sources if I wanted further information. About 90 percent of what HeadsUp gave me was exactly the kind of news I wanted to keep tabs on, whereas the Executive News Service on CompuServe, which I also used, picked up way too many extraneous stories, such as news of global politics. However, the advantage to ENS was that it gave me full articles rather than just short teasers. To keep up with *The New York Times* and the *Wall Street Journal*, I signed up for DowVision, on the Web. Had I been interested in stock market reports and sports, I could have used the Journalist software with either CompuServe or Prodigy, and on CompuServe it would also have picked up my ENS reports. (Also see *Custom news-search services*, at the end of this chapter.)

**12.** Scan daily headlines or news summaries. Most of the commercial services also carry reports from the major newswires. AP Online on CompuServe (Associated Press, GO apo), for instance, is updated every half hour, and you can choose from national or international stories, features, health and science stories, and several other categories. These are many of the same lead stories you'll hear on the evening news or see in the next day's paper, but you can get them earlier.

**13.** Search TV listings. Because the same company that owns Delphi owns *TV Guide*, watch for these listings to show up on Delphi in late 1995. If you're browsing the Web, you might want to check What's On Tonight, which is organized by regions of the U.S. and by categories of shows. Either can be quicker than reading through the listings in your Sunday newspaper's TV insert, although the main value of checking TV listings online is being able to search by several criteria, such as only movies, or only Bette Davis movies, or only concerts. You can't do that yet, but it's coming. The other advantage is that these listings will be more up-to-date than those in any printed form.

**14.** Search movie reviews. Not only can you read Roger Ebert's review for free, you can correspond directly with Roger Ebert on CompuServe. You can also search Magill's Survey of Cinema there, but it will cost you extra; on

Prodigy it doesn't. Just about every large commercial service has some kind of movie database, and many small services have local listings. On the Web, try the *Playboy* magazine movie reviews and Cardiff's Movie Database Browser, a.k.a. The Internet Movie Database, which is a must for movie trivia buffs.

*15.* Search restaurant guides. Whether you're planning a trip to another city or just a special night out, you can make it an adventure by searching Zagat's and other restaurant guides on the commercial services. Fair warning, though: Some services charge to search the same guide that's available for free on another service.

*16.* Pay bills by modem. Unless Microsoft Network enters this arena too, the only service that offers even limited online banking so far is Prodigy. But their BillPay service (see Fig. 12-3) is basically the same thing as CheckFree, an independent service that enables you to pay bills by modem. Many local banks offer similar services, but they typically cost more than CheckFree. See chapter 16, *Managing Your Money by Modem.*

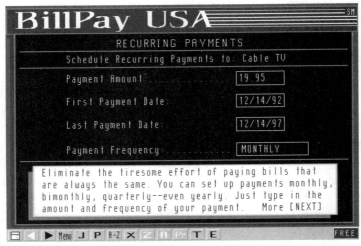

**12-3** *Prodigy's BillPay service, which is much like the independent service, CheckFree.*

*17.* Search Books in Print or the Library of Congress. If you need only one citation, it's probably smarter to call your local library's reference desk. Books in Print on CompuServe charges $2 to search by author or title, then another $2 if you want the full citation. So you blow four bucks just to find out who wrote a book or when it was published. That's not worth it, even when you're in a hurry. However, if you want to know everything in print by a certain author or on a given subject, it's far faster and less expensive to use the gopher tool to get to marvel.local.gov 70 or telnet to LOCIS and search the Library of Congress catalog, even if you live in Washington, D.C.

*18.* Order books at a discount. Do a Web search using *books* as a search term in either Yahoo or the WebCrawler search engines, and you'll find links to places like Book Stacks Unlimited, with 240,000 titles; The Antiquarian Booksellers Association; Wordsworth Books in Boston; Powell's Technical Books and the Computer Literacy Bookstore; and Books That Work, for how-to books, among several others. CompuServe and some of the other commercial services are good places to order computer books and bestsellers, sometimes at a 10% to 20% discount, but their selections are more narrow than stores on the Internet. Beware that they usually don't tell you about the shipping charges, which can negate the discount, but that's true of mail-order book clubs too.

*19.* Find a good daycare center or babysitter. The best way to do this is to join a local BBS or look for a local section on your city on the Internet or on one of the commercial services that are creating regional sections, such as Women's Wire. The forums, sections, and newsgroups for parents are also good possibilities. In all cases, be sure to get acquainted with the regulars there first, then ask people you trust, by e-mail, for recommendations. It's okay to post a public message if the participants are all from the same locale and thus would be likely to know local people, but exercise the same kind of caution and judgment you would in taking recommendations from a group of people at a public meeting offline, and don't post details about your child or your neighborhood. Sections specifically for working women or working mothers are also likely places to look. Again, bear in mind that not everybody on a large national service wants to hear about your search for a babysitter in Louisville. Local services, however, can be great resources.

*20.* Find a new job. Yes, it is possible to get a job offer online, even one in a different city. It has happened to me and thousands of others. But people find jobs mostly by networking and establishing connections and credibility with other people online, where people in their fields congregate, just as they do in the real world. For more about this, see chapter 13, *Career guidance.*

*21.* Look for a new home to buy or rent. Realtors are well-represented online already, both on the Internet and the commercial services, so use *real estate* or *house* as a keyword, and just follow where that leads. For more information, see chapter 7, *Domestic domain.*

*22.* Make relocating easier. If you're relocating, filling out the HomeFinders questionnaire on either CompuServe or Prodigy is a good place to start. They'll have a relocation counselor call you and send a Community InfoPak with discounts to other services related to HomeFinders, all for about six bucks. Their goal, of course, is to match you with one of their real estate agents, which might be better than trying to find one in an unfamiliar city yourself. But the next thing to do is post a public message in a forum you frequent, asking for guidance about good neighborhoods, schools, taxes, and so forth from people who live where you'll be moving, and in the business-related forums if you're looking for a job in that city.

*23.* Download tax forms. Among the commercial services, CompuServe has an advantage here because they're owned by H&R Block, the tax-preparation people. But other tax companies, as well as the IRS itself, are on the Web, and most of them stock the most commonly used and some not-so-common tax forms that you can download, fill in on your computer, and use to file electronic returns. Just think: Never again will you have to stand in a long line the week taxes are due because you're minus one form required for your return. For more information, see chapter 16, *Managing your money by modem.*

*24.* File your taxes. You can also file your tax return by modem through these same services, which means the IRS can process it faster, and that means you get your refund in three weeks rather than eight weeks. If you use one of the popular tax software programs, such as Turbo Tax or Tax Cut, you can file your return electronically without being on any online service, as long as you have a modem.

**25.** Shop for groceries and have them delivered. Online grocery services are rather new, and most start with a limited geographical distribution, but they'll steadily branch out. Shopper's Express, on America Online, delivers from neighborhood pharmacies as well as supermarkets if they have an agreement with ones in your area. To find out, go to Shopper's Express and type in your zip code. The Smart Food Co-op on the Web (see Fig. 12-4) is a buy-in-bulk traditional food co-op based in Cambridge, Massachusetts, but they intend to expand their service area. Peapod, based in Evanston, Illinois, operates independently there and elsewhere, but also works by modem. Expect more to follow an idea whose time has come.

**12-4** *The Smart Food Co-op on the World Wide Web.*

**26.** Find a good business or computer consultant. People in the business and professional forums will be able to recommend consultants they've worked with for almost any conceivable project. If you post a public message asking for recommendations, you'll also get e-mail from consultants who might not be qualified, so rely on references and recommendations, just as you would offline. If you have a special computer project, go to the forum or newsgroup that specializes in that kind of software or function, or ask in one of the computer forums which would be the best one to try.

**27.** Read sample issues and subscribe to magazines. The Internet Newsstand and Hearst's Multimedia Newsstand, on the Web, are good places to browse articles in magazines, the same way you would at a newsstand in real life, to see what appeals to you. They hope you'll sign up for a subscription, of course, as do all the magazines available everywhere else online that try to make it as easy as point-and-click for you to say "bill me."

**28.** Check your daily horoscope. You can download complicated software, either shareware or public-domain, to plot your own horoscope or use the online version on CompuServe (which will even help you figure out your latitude and longitude, but you must know the hour of your birth), or just read one of the horoscopes on the commercial services (and on the Web, eventually) such as the daily one by Jeanne Dixon on Prodigy. If you typed in your birthdate when you signed up, it will automatically give you the horoscope for your sign for that day, and you can put it in Favorite Places, so it takes only seconds to check it when you log on to get your e-mail.

**29.** Order gifts to be delivered. On the commercial services, it's easy to find boutiques and specialty shops because they're all in whatever they call their online malls. The Web has malls too, although it's more like a huge town, with several malls of all sizes, as well as all kinds of shops everywhere. So it's harder to find something unless you know what you're looking for and can use a keyword to search for it, yet it's easy to find gifts you might not have thought of by just typing in words such as *gift* or *shopping* or *food* into Yahoo or another search tool on the Web (see appendix G for information on finding people and places on the Internet, and also chapter 16, *Managing your money by modem*).

*30.* Send a document by modem rather than overnight courier. For less than the price of postage and far less than overnight express charges, you can send any straight-text document and most graphics files to anyone else with a modem or an online account—even if it's 30 pages or more and even if it's going to a country on the other side of the globe—and it will arrive almost instantly. If the person receiving the document uses the same word-processing software, then it might not need to be text-only.

*31.* Determine whether what you inherited is valuable. Collector's forums abound, and there's a section for almost anything anyone ever thought of collecting. Many of these people are experts or can tell you where to find someone knowledgeable and reputable to appraise that vase or coin collection Aunt Alice left you. For more information, see chapter 10, *Pleasurable pursuits.*

*32.* Retrieve rare or exotic recipes. There was a time that the people who designed the online services (mostly men) thought online shopping and recipe databases were what would be most appealing to women, and that those would lure more women online. Those are the same people who think we read only *Good Housekeeping* and *Better Homes and Gardens.* Only partly because there are still more men online than women, the people discussing recipes in the cooking forums are just as often men as they are women. So much for stereotypes, once again. For more information, see chapter 7, *Domestic domain.*

*33.* Choose a house plan. You don't have to hire an architect, settle for a house that looks just like the one across the street, or even spend time and money on magazines to have your choice of many house plans. Just check the *Home* magazine section on America Online or use *house* as a search term on the Web to find other sources. You can browse through them to your heart's content, and order any you want. For a sample, see chapter 7, *Domestic domain.*

*34.* Take a class. There are experimental classes on the Internet, everything from leisure courses to professional licensing exam preparation on GEnie, and a growing array of inexpensive courses from genealogy to how to navigate the Internet on America Online, plus freelance journalism and PR classes on

CompuServe. They usually cost less than local adult education classes, and the classes actually meet online once a week, with assignments in the interim.

**35.** Work toward a college degree. You can get a master's in journalism or business on CompuServe, take classes through the Electronic University in America Online's Education center, or sign up directly with many colleges and universities to take the classes they offer by modem and other forms of electronic communication, all under the rubric of distance learning. For details, see chapter 9, *Lifelong learning.*

**36.** Check weather and ski-condition reports. There are some weather services online that are fun to look at even if you don't understand them. The ones from the University of Wisconsin and Purdue University are particularly good, both on the Web. Purdue has satellite maps from the National Weather Service and a 24-hour forecast (see Fig. 12-5). You can also find just about any data you might want directly from the National Oceanic and Atmospheric Administration on the Web. And on WebWeather, you'll find reports for aviation, historical weather data, and regional reports, although those reports aren't as current. On CompuServe, you can get a report just for your city. If you want to know only whether to carry an umbrella tomorrow, stick with the nightly

**12-5** *The Purdue University Weather Processor on the Web, in cooperation with the University of Wisconsin and the National Weather Service. This shot shows a storm front approaching that brought floods to California.*

news on TV. But if you're planning a vacation or any kind of trip involving out-door sports, you can find information online faster than waiting for the cable TV weather channel to scroll to just what you want to know. There are also specialized reports for skiers and seagoing folk, and some travelers say they find the aviation weather reports to be the most reliable and useful.

*37.* Calculate loan or mortgage payments. You'll find mortgage and loan cal-culators many places online. Check the personal finance and real estate forum libraries for software you can download, and some of the real estate services on the Web, such as the Home Buyer's Fair. Using *mortgage* or *loan* as a keyword will also turn up a quick-and-easy version to use online on some of the com-mercial services. You can even apply for a mortgage from reputable institutions online, mainly on the Web.

*38.* Join a support group. Although you can't see the people in an online sup-port group and give them real hugs, you aren't likely to run into them at the supermarket either, so the relative anonymity makes it easier for many to be candid. People definitely do develop strong, genuine bonds online, and this is one of the better uses of the online medium. Internet mailing lists and the health forums on commercial services are the best places to look, but people often develop their own support groups too, and meet at a scheduled time in a private chat room they set up just for that purpose. See chapter 11, *Psyche and spirit*, and chapter 8, *In sickness and in health*, for more leads.

*39.* Find a long-lost friend or relative. There are numerous ways to find every-one from birth parents to long-lost college roommates online, some obvious, most not. The most obvious is to pay $18 to a service on the Web called Find-a-Friend that searches various databases by last-known address, maiden name, surname, neighbors, or Social Security number (good for debts, they say). For more information on a few of the resourceful methods people have devised and used successfully, see chapter 10, *Pleasurable pursuits*.

*40.* Study genealogy and research your family tree. Most commercial nets and the Internet have at least a section if not a whole forum or newsgroup on gene-ology, and some, such as America Online, have formal classes too. Because you can get help from people all over the world and because it's such a natural place to learn how to do research, this is one of the more popular topics online.

*41.* Manage your investments. You can buy, sell, trade, do research, get and give advice, and just generally have more fun using the many money management tools, services, and discussion forums online. For much more on this, see chapter 16, *Managing your money by modem.*

*42.* Join a special-interest discussion group. You name it, there's probably a forum, a section within a forum, an Internet Mailing list, or a newsgroup for it already; if there's not, you can eventually start one once you learn the ropes. From hobbies to business to babies, the forums on the commercial services and the newsgroups and specialized mailing lists on the Internet are among the richest resources online, and the most popular after e-mail.

*43.* Circulate your résumé. You can upload your résumé to the library of a business-related forum on the commercial services or to an Internet newsgroup, as well as to one of the many career databases on the Web. It's usually free so it's worth a shot, but don't expect much more in the way of results than you'd get from a shotgun mailing in the real world. It's better, as it always is, to send it to people you meet online who are actually in a position to hire or recommend you or tell you about openings. For more on how to build a job-search network online, read chapter 13, *Career guidance.*

*44.* See if someone else is using the business name you've chosen. Both GEnie and CompuServe offer searchable databases for trademarks and trade names, which are ostensibly the same government database for national registrations that lawyers use. They cost less on GEnie, and you'll still need to check local and state sources, but it can save you a lot in lawyer's fees. The federal patent registration database is accessible through the Internet. For further information, see chapter 14, *Business management and marketing.*

*45.* Find out how your senator or representatives voted and what they're voting on. It isn't the only option, but one of the best for voting records of federal elected officials is Cap Web on the Web (see Fig. 12-6). For pending legislation directly from the U.S. Congress, stop by Thomas, the site on the Web named after Thomas Jefferson and created to disseminate such information.

**12-6**
*Cap Web, the "unofficial guide to Congress," which contains voting records of U.S. senators and members of the House of Representatives, and guidelines for lobbying your representatives.*

**46.** Send letters to Congress. Check any of the politics forums or newsgroups for more information, FIND congress on CompuServe, or do a Yahoo search using the terms *congress* and *addresses*, and you'll find multiple options, including e-mail addresses for all members of the U.S. House of Representatives and the Senate who have them so far.

**47.** Get voter guides for federal elections. The League of Women Voters is in the Political Debate Forum on CompuServe, and you're likely to find guides from them, as well as propaganda from major special-interest groups anywhere and everywhere online they think they can reach voters as any major election draws near.

**48.** Teach your kids and explore new places and ideas together. The World Wide Web, especially, is a great place to go exploring with children from about five years or older. Even some of the problem-solving games online or interactive areas such as the ones Prodigy offers are good ways to spend time together and get to know how your children think, and guide them in decision-making without lecturing.

**49.** Choose the right college. Peterson's College Database and Peterson's Graduate School Database are accessible through CompuServe, GEnie, and the Dialog research service, and there's helpful information available through Peterson's Education Center on the Web. The databases are searchable, so they save a great deal of time. There's also financial-aid information at the Peterson's Web site, on eWorld, and in some of the education or student forum libraries on the commercial nets.

*50.* Look up details about prescription drugs. Look for *Consumer Reports'* Complete Drug Reference on the commercial services and PharmNet on the Internet. For more information, see chapter 8, *In sickness and in health.*

*51.* Extend your professional network. There's no better way to do this, online or off, than through the business-related forums on the commercial services (CompuServe has the most and the best for this purpose) and the Internet mailing lists. The newsgroups on the Internet are also an option, but the mailing lists are more focused and useful if you can find the right one. See chapter 13, *Career guidance,* for information about some of the forums.

*52.* Attend events featuring politicians, celebrities, and experts. Bill Clinton, Mick Jagger, Edward Albee, Barbara Walters, Harlan Ellison, Tom Clancy—the list of such people who have been guests at free public conferences on America Online, CompuServe, and Prodigy, particularly, is long and impressive. Such events have become routine, and all you have to do to attend is log on. You might not be among the few who get a typed reply to your question, because there will be many vying for that privilege, but that's no different than attending a similar event offline.

*53.* Adopt or name a pet. People who hang out in the many pet-lovers forums and newsgroups are nothing short of fanatical, and always knowledgeable. They delight in helping people solve any kind of pet-related problem, or even in helping you find one to adopt or helping you name one you already have. Many breeders and pet-rescue network volunteers participate in these discussion areas, so it's a good way to find a pet whose history you know something about.

*54.* Explore far-away places. You can visit a museum or an art exhibition across the country or halfway around the world, or tour the Vatican's Renaissance collection, or communicate with the astronauts aboard the space shuttle through NASA's site on the Internet or explorers on a Mayan or South Pole expedition on Prodigy (see Fig. 12-7) or . . . well, you get the idea.

**12-7** *During the Maya Quest expedition, you could communicate with the explorers through their section on Prodigy.*

**55.** Check train schedules. These are cleverly hidden under the title Train Schedules on the Web. Type that into Yahoo or WebCrawler's search form, and you're on the right track.

**56.** Find subway maps for major cities worldwide. Type subway into one of the search engines, which will lead you to a menu-driven service with maps for subways in what must be all major cities in the world that have subways, because it covers about 15 cities. You type in the street coordinates of where you want to start from and where you want to go, and it will tell you what subway trains to take, show you any connecting trains you have to catch, and estimate how long it will take you to get there. If you add 20 minutes to the time estimate, it's a pretty cool tool, both for locals and for tourists.

**57.** Create a Web page for your business or your family. "Everybody who's anybody will have their own Web page eventually," says Eugenie McGuire, a multimedia developer who, sure enough, has her own Web page (or site). I'm hoping she's wrong, but the way it's beginning to look she might not be. There are more than 4,000 people or families with personal Web pages telling about themselves so far. For an example of one of the better ones, and why and how they did it, see the section about the Farmer family's home page near the end of chapter 10, *Pleasurable pursuits*, plus comments from the pros in chapter 15, *Online marketing methods*.

*58.* Enter a contest. Business sponsors on both the Web and the commercial nets run a steady stream of contests, mainly as advertising gimmicks, marketing tests, and to build prospect databases, which are the same reasons they do them in any other media, of course. To find the latest list on the Web, use *contests* as a search term in Yahoo.

*59.* Plan a wedding. Check any of the software or shareware libraries on the Net or the computer forums of the commercial services for bargains on wedding planning software. Basically, they're preconfigured databases, but can make what can seem like a formidable task a lot easier. There's also a newsgroup or two on the Net about planning weddings (check the soc., alt., or misc. hierarchies), and sites set up for people to chronicle their own weddings, complete with photos and more narrative details than you'd probably ever want to know.

*60.* Research the law. Although there's a Law Forum on CompuServe and many databases or lawyer sites on the Web, the best place to start is the Law section of The World Wide Web Virtual Library, because it has pointers to just about everything else on the Web.

*61.* Shop for a new car. You'll find automobile or car forums or newsgroups for car fanatics to discuss details to their hearts' content, from the latest models to classics or collector's models, as well as an increasing number of Web sites and information storefronts operated by dealers. There are also databases to search for the latest specifications and sticker prices, and places to check the blue-book value for used cars. The best place to start, however, is Consumer Reports, which is on CompuServe, America Online, and elsewhere, although it's set up differently in each place. It gives you a lot of information, all of which is useful and can save a lot of thumbing through pages at the library. (See Fig. 12-8.)

**12-8** *Corvette advertisement and information section on Prodigy, complete with photo option.*

*62.* Get free software. Except for Prodigy, which heretofore had little in the way of libraries attached to forums, every computer forum on the major services has a library stocked with public-domain software, try-before-you-buy shareware, or working demos of commercial programs. You'll also find software tucked away in libraries for almost any kind of forum, available by FTP from many places on the Net. The most popular free screen saver in one of the children's forums we checked is one that makes spiders crawl across the monitor screen whenever you're idle. See chapter 7, *Domestic domain*, for more appealing examples.

*63.* Check a company's credit history. It will cost you around 30 bucks, maybe more, but there are several services online that will run credit checks for you on a company that you're thinking of doing business with, but want to check out first. You can do this through Dun & Bradstreet offline too, so it's a matter of which is more convenient and which costs less for comparable results. Check the general reference and business sections.

*64.* Order almost anything, including a kitchen sink, to be delivered. Yes, you really can order a new kitchen sink online (on the Net, at least), or lumber, or a garden tractor, or furniture made from twigs for that matter. Or something more mundane, like flowers, or fruit, or a music CD. You can even order the CD from a CD. America Online's "to market" CD lets you watch a video clip or hear a sound clip of an album before you order it—assuming you have a multimedia-equipped computer, that is. For the requirements, see appendix A.

*65.* Consult garden and plant guides. Whole gardening encyclopedias are online as searchable databases, and gardening forums and newsgroups are as common as dandelions. Whether it's a question about how to save your withering houseplant, designing a formal garden, or what's the best ground cover for the location and exposure of the new house you just bought, you'll find knowledgeable, helpful people, and probably get an authoritative answer to any question within 24 hours, which might be in time to save the plant. Also see chapter 7, *Domestic domain.*

*66.* Tour an art exhibit. The WebMuseum, shown in Fig. 12-9, is another one of the most popular Web sites. Created by a Parisian, it has photos of paintings from the Louvre, along with catalog descriptions that tell about them, a rare Medieval art collection no longer open to the public for in-person visits, and more to come. There are also art exhibits sponsored by private and university galleries, as well as museums worldwide on the Web. To find a few of them, start at the Expo site or use *painting, art, gallery,* or such as search words in Yahoo.

*67.* Visit a natural history museum. One of the most popular sites on the Web is the Paleontology Museum at the University of California, Berkeley. It's one of several worthwhile places you'll be able to see at least part of, through photos and narrative guides, on the Web. *Hint:* Check the Expo Web site at:

http://sunsite.unc.edu/expo/ticket_office.html

*68.* Publish a newsletter. Without so much as a postage stamp or a printer, you can publish a newsletter online, either in straight text and delivered by

*Bienvenue au WebMuseum!*

You are now at the WebMuseum Pyramide-guichet, near the world map cabinet. Before you start your journey, please note that most documents have inlined images, the size of a thumbnail. To view the full-sized versions, just follow the hyperlink (just click on the embedded image usually).

 Welcome to the WebMuseum!

I wish you the most pleasant visit.

**12-9** *The WebMuseum, and art gallery on the World Wide Web.*

group e-mail or, if it's on the Web, with full-color graphics and photos. There's software to create documents for the Web that does the work for you, so you don't have to learn the programming codes. Some of it is even integrated in the best-selling word processing programs, or soon will be. As one who publishes an online newsletter, I must warn you that it's a lot more complicated than I'm making it sound, both in the publishing and the marketing, but if you want to publish either a free or subscription online newsletter for your business or for an organization you're involved with, you now can, and for far less than it once cost.

**69.** Find photos or graphics for your club's newsletter. If you're going to do the newsletter, you need art and photos, right? There's plenty of free clip art online, and even high-quality, rather generic photos that you can use for free if it's not for a commercial or for-profit purpose. If it is, each use will cost $250 or more, and must be licensed or authorized in writing, or you'll be violating a copyright. Make sure you understand what's public domain, or free, and what's not, because this is federal and international law. The files are usually clearly marked, however.

**70.** Send press releases. On the Internet, you can find an e-mail list for the media that's, supposedly, updated regularly. It's often available within Internet forums on the commercial nets too. For guidelines on what works and what turns the media off, see chapter 14, *Business management and marketing.*

*71.* Tune into a band. The Ultimate Band List, which is part of the World Wide Web of Music (on the Web, of course), has 2,770 links to a thousand bands with more than 800 Web pages of their own. Whether you're a fan or want to book a band for an event, you'll find everything from classical to industrial rock right here.

*72.* Discover new artists. Both visual and performing artists are using the Web very resourcefully and imaginatively, and it turns out to be a great medium for people to get wide exposure who might not otherwise because of the usual agent-gallery game. Either independently or as collectives, they're creating thematic exhibits—some of which invite visitors to leave messages or add to the creations—or showcasing collections of their works.

*73.* Participate in a national opinion poll. As you've probably already seen on the evening news, the media and businesses now frequently poll people online on controversial issues, politics, or attitudes about certain topics, and report the consensus. The results are terribly skewed because of the demographics online—largely white, well-educated, urban men in their 40s earning good money—but it's another one of those tidbits that's good for slow news days.

*74.* Estimate your Social Security earnings. Commerce Net seems a strange place to find the Social Security Administration on the Web (see Fig. 12-10), but that's where they are. Most of the information you'd have to request by mail and wait weeks to get is there, including charts to help determine what your retirement income from Social Security might be. However, you'll still have to get the report on their records of your earnings the old-fashioned way, unless you've kept a careful tally over the years yourself.

*75.* Write to friends, relatives, and colleagues worldwide. E-mail is everybody's favorite feature on any online service because they can write quick notes to people they wouldn't have time to keep up with otherwise, at any hour that's convenient for them or you, and save considerable money on long-distance phone charges. No matter where they are, the recipients get the mail almost instantly, and it's much easier and quicker than messing with paper and stamps. See chapter 4, *Friends and family,* for creative ways people are using e-mail.

*SSA ONLINE*

Welcome to **SSA Online**, a service of the Social Security Administration, with headquarters in Baltimore, Maryland, USA. Information on this server is maintained by the SSA Online Team **e-mail:** <u>webmaster@ssa.gov</u>.

**12-10** *The Social Security Administration's site on the Web, at Commerce Net.*

*76.* Calculate currency conversion rates. If you're headed out on an international business trip or vacation, stop by TravelWeb or the Global Network Navigator on the Web and check the currency conversion rates. Print out a report and tuck it in a side pocket of your bag, and you're on your way.

*77.* Get help with homework. America Online, GEnie, Prodigy, and some of the other major commercial nets have homework-helper or tutoring services, sometimes including access to real people who are generally teachers themselves. You pay either regular online rates or a monthly charge for a specified number of hours so your child can have access to this service. See chapter 9, *Lifelong learning*, for details.

*78.* Get to know people in other countries. On any of the services that are international—and most of them are or soon will be, although CompuServe, Microsoft Network, and the Internet are the ones with the farthest reach—you'll meet people from other countries in any of the forums. If you want to practice a foreign language, however, there are forums and newsgroups where people communicate in native tongues. France, Germany, England, Scandinavia, Australia, and Japan are especially well-represented online.

*79.* Get in-depth health information. Everybody is interested in health to some degree, so there's an abundance of health information everywhere online, from research databases to discussion forums and support groups. See chapter 8, *In sickness and in health*, and chapter 18, *Online research basics*, for tips on how to find both people and information.

***80.*** Find kindred spirits you'd never meet otherwise. If you're one of the few people in your community who's a true Trekkie or you know nobody else who's interested in a particular religion or hobby, you'll wonder how you ever got along without being online because you'll find hundreds of kindred spirits. You might have to look for them, but I can almost guarantee you that they'll be there, somewhere. Just ask around in likely places.

***81.*** Buy or sell used items. Some forums have classified ads, and there are sections for various categories of classified ads on every online service. If you have something unusual to sell or that you want to buy, and don't mind shipping charges, you can advertise for free or for less than newspaper rates. What sells best, though, is computer hardware and software, unless your ad is within a forum where people most likely to be interested hang out.

***82.*** Lobby for your favorite cause. Because it's such a mental medium, people online place a high value on well-reasoned arguments, and they love to debate and discuss. Yes, there are flamers, or rude, hostile, irrational people, but they're not the norm. Many people say one of the things they like about being online is that it exposes them to ideas and people who think differently from them, and that they sometimes change their minds or modify their stances a bit as a result. So if you can present your case clearly and cogently, you'll have an automatic audience, and you just might persuade a few of them to see it your way or to support your cause by votes or dollars.

***83.*** Volunteer or make a donation. Many U.S. and international nonprofit organizations are online, and most of them will be within the next year or three. The Internet Nonprofit Center and others are actively recruiting volunteers online, knowing that they're reaching an audience that has more money to spare than time, usually. So they've set up ways to make contributions, and provide good information on how to screen which causes are worthy of your support. See chapter 17, *Direct democracy and community involvement*, for pointers to some, or start with The Internet Nonprofit Center, America's Charities, or ReliefNet on the Web. Each has links that will lead you to other organizations, although there's currently much more information about health and social service organizations than the arts.

*84.* Bone up on global politics and geography. There are numerous forums and newsgroups where people discuss global politics, often with people from the countries that are involved participating, and numerous ways to do further research, including online editions of newspapers and their archives, research databases, and maps (see Fig. 12-11).

**12-11** *Map of Cherchen from Magellan Maps on CompuServe.*

*85.* Send a Federal Express or UPS package or track one. Federal Express is on the Web, and both companies are on a couple of the commercial nets, although none of the Federal Express customer service reps know what you're talking about when you call to say your attempt at placing a pick-up order online didn't work. They're new at this, but they'll catch up eventually. Maybe. Whether this is easier or quicker than just calling the usual number is questionable. However, when it's fully functional so you can track a package gone astray without having to order their software, that might help.

**86.** Check your American Express account. If you're a cardholder, you can now check the status of your account on their ExpressNet service on America Online, plus pay your bill, make travel reservations, and take advantage of their special offers. They'll gladly take your application for an American Express card too, of course.

**87.** Plan your diet. Check any of the health or food-related forums' libraries, and you'll find several software programs for counting calories, or any other way you want to diet, as well as charts that tell you how much you need to lose and tests to take to determine how fit you are or aren't.

**88.** Meet a date or mate. Although the odds are against it, many people have dated or married people they met online. Maybe it's because we live in the era of AIDS, but just as many people seem to enjoy the online romances with people they never meet as much as trying to turn fantasies into reality. See chapter 6, *Sex and romance*, for all the juicy details. Well, some of them anyway.

**89.** Find a mentor. Traditionally, you don't find mentors; they find and adopt you. But it can work the other way around online, and even if the relationship with your online mentors aren't as direct and personal because you seldom if ever see them, they can be fulfilling and helpful. See chapter 13, *Career guidance*, for how to connect with potential mentors online.

**90.** Join a book discussion group. Great Books discussion groups and similar reading groups seem to have all but disappeared in the real world, but they're flourishing online. You don't have to show up at a real meeting, which is one of the reasons people like them. The discussions continue every day, so you can participate when it's convenient for you to read what people have posted and reply. You'll find the groups in corners of the forums on literature or sometimes in the ones for writers, as well as among Internet newsgroups and mailing lists.

**91.** Keep up on international developments in your field. There's no faster grapevine than the Internet and some of the business forums on CompuServe, so even if you have time only to lurk and check message headers, you can get wind of what's happening as it happens if you find a forum, newsgroup, or

mailing list related to your business. There are also many trade magazines and
newsletters in the research databases, so you can search once a week or once a
month on a topic that you need to follow, and keep abreast of developments.

**92.** Scope out a competitor. With the specialized services such as Dow-Jones
News Retrieval, DowVision, and Nexis, the various company and stock reports,
the research databases with the trade publications and financial reports, and
the gossip among knowledgeable people in the forums, newsgroups, and mailing lists, you have more tools at your disposal to find out what others in your
field are doing more easily and faster than ever before, all without hiring a
research staff or leaving your desk. With customized news or analysis, such as
Headsup, via e-mail, Infoseek on the Web, or @Advantage on Interchange, you
can get daily updates or summaries in your e-mailbox, and cut your reading
time and that stack of publications you never get around to reading anyway
down to almost nothing, yet still be able to talk about the latest news at the
breakfast meeting. See chapters 14 and 15 for comments from some women
on how they stay on top.

**93.** Hold a meeting. Through Internet Relay Chat (IRC) on the Internet and
chat areas on the commercial services, you and anyone else you choose can set
up a private chat area. Even though you'll pay by the minute, it usually costs
less than long-distance phone charges, so it's a good way to hold a business
meeting or get together weekly with friends or relatives. Just tell everyone
ahead of time where and when to log on.

**94.** Locate a zip code. On Commerce Net, on the Web, you'll find a map of
the U.S. that you can search by zip code or for zip codes if you know only the
name of the city. Most cities have more than one zip code, however, so that's a
potential problem. It will also give you latitude and longitude, elevation, population, and other data, plus local and regional maps.

**95.** Play games with other people at any hour. Use *games* as a keyword on
any of the commercial services or try The Games Domain on the Internet. Be
prepared for a lot of stuff based on fantasies or violence, but there's more than
that. *Warning*: The MUD (multiuser dimensions or dungeons) games on the

Internet become genuinely addictive for some people, so proceed with caution if you have obsessive tendencies. Seriously.

**96.** Reduce your bills for long-distance phone calls. The answer here is e-mail (or the chat sessions mentioned elsewhere, although you won't save nearly as much that way, and real-time chatting can be too habit-forming for the good of the budget). There's also a company that has come up with a way for people to hear one another while chatting on the Net, thus bypassing the long-distance companies altogether, but we suspect they'll find a way to stop that before this book even gets into your hands. Just in case, though, ask about it in the Internet forums on one of the commercial services or on the Net itself.

**97.** Follow your muse. On America Online, eWorld, Prodigy, the Internet, and elsewhere you'll find U.S. and international schedules for live theater, dance, concerts, and festivals. Yahoo, the Web search tool, has an especially good list for concerts and festivals, but check the libraries of any of the relevant arts forums too, and *Playbill* magazine's listing of plays in their section on Prodigy.

**98.** Find out what's free. Although you'll find a lot of free things online, the single best place to find out about freebies in general is Matthew Lesko's Information USA section on CompuServe.

**99.** Hire a translator. You could do this through any of the relevant foreign-language forums or newsgroups, and perhaps get something translated gratis by another member, if it's something short and not a great imposition. If you're really in a hurry, though, send e-mail to info@glnk.com, which is a service that charges for translations in English, Spanish, French, and German. You'll get a quick reply telling you how it works and what to do next.

**100.** Track down a quotation. On both the Internet and a couple of the commercial services, you'll find databases of quotations for those times you need something quickly for a paper or a speech. If there's a quote you have in mind, but can't quite remember how it goes or who said it, ask in a relevant forum— particularly the ones where book lovers hang out—and someone will probably tell you who said what and exactly how they said it within 24 hours.

# Custom news-search services

Most of these cost extra, but can be cost-effective, especially for business use. Those that are by subscription typically offer a one-month free trial.

## DowVision

DowVision culls and indexes same-day stories from the *Wall Street Journal*, *New York Times* News Service, and several major news wires. In conjunction with Ensemble's Relevant software, they offer a version of just the *Wall Street Journal* that comes straight into your computer.

## Executive News Service

CompuServe's Executive News Service is an electronic clipping service that monitors major newswires around the clock for information of your choosing, based on up to 72 keywords. It takes considerable fine-tuning to keep from getting news you're not interested in, and you will even then, but it's included in your monthly fee.

*Tip*: It's fastest and easiest to scan ENS manually, in terminal mode rather than with CIM. You can run through headlines quickly, send the ones you want to read to your e-mailbox by typing SEND Story# and hitting Enter. Then you can read them offline when they download with your e-mail. GO ENS.

## Farcast

Farcast is an intelligent agent- or Knowbot-based, customizable news-search service that works via e-mail for $30 a month. It covers news, sports, weather, business, and entertainment stories from the news and PR wire services, plus stock quotes. The Broadcast option collects industry-specific stories, much as HeadsUp does. For details and a demonstration, send e-mail to info@farcast.com and demo@farcast.com.

## HeadsUp

Available on eWorld via e-mail, this electronic clipping service uses Knowbots to scan newswires, press releases, and trade magazines and newsletters for business-related news, and sends headlines and short summaries to your e-mailbox every morning. The relevancy rate is very good, but you get only teasers because they want you to pay three bucks a pop for the full story plus your $30 a month subscription. So you know that something happened, but not exactly what. The consensus among news professionals is that the relatively high cost isn't worth it because too many of the stories you pay extra to get are press releases. If it were less than $10 a month, most agree the brief Headsup notices would be worth it. There's now a Web counterpart, called NewsPage, at http://www.newspage.com (see Fig. 12-12), or call 800-414-1000 or 617-273-6000 in the U.S.

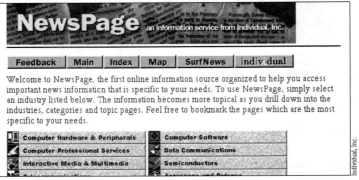

**12-12** *NewsPage on the Web, an electronic business news clipping service operated by Individual, Inc., the company that also offers HeadsUp summaries by e-mail.*

## Infoseek

The custom news part of this subscription service searches major news and PR wires and magazines and offers a high hit rate for a low cost, with flexible rates ranging from $1.95 a month for 10 searches, $9.95 for up to 100, or 20 cents per single search if you're not a subscriber. You can also use it to search all Web and Internet sites, including newsgroups, on a particular topic. On the Web at http://www.infoseek.com.

## Journalist

This is a separate software program (about $20 to $30, depending on the latest deal) that works with CompuServe and Prodigy. Once set up to follow the kinds of information and sources you want, it provides a somewhat customized personal daily newspaper covering news headlines, sports, stock quotes, and limited business news. You can get more information or order Journalist on either CompuServe or Prodigy.

## Newshound

For only five bucks a month, the *San Jose Mercury News'* Newshound service will watch for stories on topics of your choosing from their own reporters, wire services, and even from their classifieds. Reports from early subscribers are good. For a free 30-day free trial, call 800-818-NEWS in the U.S., and give them your e-mail address.

## Profound

Basically an online research service like Nexis, but somewhat less expensive, Profound is due to debut mid-1995, and there weren't many details available at press time. It will have a customizable news-search component as well. It will be available by subscription with additional charges, such as $3.95 per article downloaded.

## Relevant

The Personal Digital Newspaper that Ensemble's Relevant software creates for you comes from customized searches of the *Wall Street Journal.* If it were also able to search the *New York Times* (as was in the works) it could have been a great thing, depending on the price. But both Ensemble and DowVision, which work with Relevant, have been unable to get the *Times* to play in anybody's yard but their own. News junkies still have high hopes for it, but the future seems uncertain, and word is that after the free trial period the cost will be high too. For a sample via the Web, go to http://www.ensemble.com.

# CHAPTER 13

# Career guidance

**U**ntil you've experienced it yourself, it seems illogical that people online would feel that, even when they use their real, full names, they have the protection of anonymity, thus are willing to be more open about themselves. It's a lot like reading revealing stories about people in newspapers or magazines. You know the names of the people and perhaps intimate details of their lives, yet neither of you are ever likely to meet unless you make a special effort to, so you keep a safe distance. However, online readers can talk to the people they're reading about and the comments are coming directly from them, not through a detached reporter. Thus you feel involved in their lives. You are, in fact, and what you say to them affects their lives. Chistine Wu's story is a good example of how it works, and how you can use the forums online to get guidance in your own career.

"I do marketing materials, technical writing, and publications design," she explains, "and I get business through referrals or by networking. I'd been doing a lot more aggressive marketing, so I was going to a lot more meetings. Over three months, I met five men who asked if I'd be interested in meeting with them to discuss possible business projects.

"When I met with them, I found out that three of those five men had used the meeting as a pretense to ask me out on a date. I found that really shocking and offensive! I always dress conservatively and professionally, and act in a businesslike manner, so I didn't feel I was doing anything inappropriate, and neither did any of the people I asked who know me."

Although she's been involved with the same guy for quite a while, Christine isn't married. Yet she resorted to wearing a fake engagement ring because these encounters made her so uneasy, but she still wasn't sure what she could do differently to prevent the passes, nor how much of the problem was the men themselves, not her. Most of us were taught to be "nice girls," so we can easily identify with her reaction: "Part of my problem is that I don't want to offend anybody. I get mad and want to kick people in the shins, but I don't want to be confrontational, because one of the things about being a professional is to never lose your cool.

"I wanted to know from both men and women if they'd run into similar situations, and how they handled the problem. I had called some friends of mine in town, but I wanted some more objective input too, so I posed the question in two different forums on CompuServe."

Christine is well-known to regulars in the PR and Marketing Forum, but says she didn't post there precisely because she has a high profile among her peers there. "Even though I know many of the people in that forum, the relationships are business, not personal," she says. "As most people do with business associates, I try to maintain that professional wall, so I was afraid that the next time I'd see them, they'd know I had these concerns, and it could be awkward."

"I posted one message about it in the *Freelance Success* newsletter section in the Journalism Forum, because I knew that the people there are experienced in meeting new people constantly, and having to go out and take the risk of being taken advantage of when they're trying to do business. I also posted a message in the Working from Home Forum, in the Getting Business section. I got messages from women who'd been in situations like mine and a few messages from men who had approached women and didn't realize that they might have made them feel uncomfortable. Nobody seemed afraid to talk about it openly, and I think that was because of the anonymity."

"A lot of the women who responded said, yes, wearing a ring does help. One woman asked me why I didn't wear a wedding band instead, so someone wouldn't think I hadn't closed the deal, so to speak, and was still approachable. The men said the engagement ring would deter some of the guys, but the wedding ring would send clear signals that I was already taken. So they advised me to wear both. A couple of men said it wouldn't make any difference to some men, regardless."

> *Their feedback made me feel more confident*
> *that my attitude was the right attitude to take.*

Christine's query came as a bit of a shock to those who had believed the ads years ago that proclaimed, "You've come a long way, baby." No company could get by with an ad phrased like that today, which lulls us into thinking things have truly changed. So Christine's dilemma led to a lively discussion that went on for two or three weeks.

"I got a lot of support from many different people, men and women, telling me not to have a victim mentality, yet others who said, 'Don't blame her!' Some men suggested that I tell the prospective clients that I have no interest in mixing business with my personal life and that's my policy, but if I was going to brush a guy off, always do it in a professional manner and don't make it a personal thing. So I now say, 'My business policy, like most major corporations, is not to date current or prospective clients.' If that doesn't work, I say, 'If you don't have business to talk about, I don't think we have anything to talk about. Nice meeting you,' and walk out the door."

In case you're wondering, yes, she still wears the fake engagement ring to meetings. Because of the advice she got online, she's thinking about buying a wedding band to go with it as a more convincing deterrent. As she says, "That's a sad commentary on our society."

Nevertheless, Christine is glad she asked for reactions online. "The advice that I got online was helpful in a different way from that of my friends. Friends tended to deal with the problem as though they were the ones examining me, giving me their personal reactions, whereas the people online didn't know me well or at all. So they responded from their own experience and said, 'This is what worked for me.' Their feedback made me feel more confident that my attitude was the right attitude to take, and that it's not worth it to put up with this in the hope of getting a new client."

# How to find mentors online

In the real world you don't find mentors; they adopt you. That's somewhat true online too, but much less so, because you gain access to people with knowledge, experience, and prestigious positions whom you might never meet if you weren't already moving in their circles. The percentage of businesspeople online is still relatively low, so those who are there are in the vanguard. They're typically sharp, successful, and leaders in their fields. And because you can meet people from all over the world, yet not get bogged down in office politics and innuendo, it's far easier for women to get objective guidance and find many mentors. Says Mary Anne Graf, a business owner whom you'll hear more from in another chapter, "There are some real authorities in the forums, people who are running multimillion-dollar companies who are willing to share what they know."

If you notice people whose comments you always watch for, and find yourself wishing you could talk with them, do. Start by replying to one of their public messages. Don't make it fawning, but it's fine to convey your respect or admiration. Ask them a question, either publicly or by private e-mail, to open a dialogue. Let your experience and intuition guide you from there to gauge whether they'd be receptive to occasional messages. You'll probably be surprised at how much more quickly people online get past the small-talk stage, and how candid and remarkably helpful they are. It's bad form not to reply to e-mail, as long as it's polite, so you'll get a response one way or the other. Just don't deluge them with e-mail if they're cordial, because they're probably having a tough time reading all they get each day already. But you needn't be a sycophant nor self-effacing either. One of the great things about communicating online is that it's much more peer-to-peer, even if your respective titles or places in the pecking order would normally create a barrier in other circumstances.

> *There are some real authorities in the forums,*
> *people who are running multimillion-dollar companies*
> *who are willing to share what they know.*

Pardon me for pounding away on this point, but it's important to become a participant in forums, not just an observer. Barbara Bennett, who owns Sarasota Occupational Therapy in Sarasota, Florida, echoes what many women say in talking about her discussions with doctors and other medical professionals in the Medsig Forum on CompuServe. They've not only helped her make crucial business decisions, but her active involvement in the forum has had an unexpected effect. "It has increased my confidence, especially in talking with doctors and the other people I associate with there," she says, "and that has spilled over to how I deal with people locally, offline."

# Beyond the obvious routes for online job searches

One of the most obvious ways to use online services in your career is to find a better job or to find one in another city, whether you're relocating by choice or necessity. But the best way to do that, as with many things online, isn't the most obvious way. Virtually all online networks have classified ads sections, and there are an increasing number of places to upload your resumé or read the classifieds in other cities' dailies or national newspapers, such as the *New York Times*. Better yet, recruiters are setting up shop on the Web, and there are one-stop shopping sites that could turn out to be the best resource, such as Career Mosaic and The Career Center on the Web. Because this is still a new medium for most of these sponsors, and the high-tech ones were the first to arrive, the job listings have been sparse and too technology-oriented. That's beginning to change.

But you probably learned long ago that 80 percent of the good jobs never show up in the classifieds or in employment agencies' databases. There are two ways people have always used to find good jobs. Just as you do offline, when you want to make a career move, you need to get connected and get noticed. But being online makes both of those things easier than ever.

Says Clara Horvath, a career counselor and out-placement consultant who serves as sysop for the Career Coach section on Women's Wire, "I find it's most helpful to extend your networking circle. People post questions like, 'I live in San Jose and have a good friend who's about to move here from New York. This is what she's interested in. How should she get started?' or 'I'm about to have a telephone interview, and I've never done one. How should I handle it?' or 'I'm going for an interview, and this is my situation with my boss [not good]. I'm worried about the kind of recommendation I'm going to get, but this other person really likes me. So how should I play it?' If you want to explore options in another community, there are people who can help you with that. I've been surprised at how much people are willing to share personal resources. They give names and phone numbers of friends and associates to contact."

Women's Wire now has sections for people according to what region of the country they live in, throughout the United States and Canada, and other services are setting up sections for regional interests too. Because of the sense of community that develops online, if you're there it's like you're a member of the club, so it's the norm for members to help other members, especially once they had a chance to get acquainted through exchanges in public discussions. As girls, most of us were taught not to interrupt and that, to be polite, we should wait until we're invited or given an opening to speak. Partly for that reason, far more women than men lurk, meaning merely read messages but rarely reply. Online, that doesn't work at all. People aren't likely to invite you into the circle because they can't see you and don't even know you're there. So all discussions are fair game and open to everyone, and everybody expects you to just jump in. It does pay off to lurk for a while when you're new to a forum so you can get a feel for the tone and topics, and gauge what's appropriate and what's not. After that orientation period, however, speak up.

If you explore and look for forums where people discuss your particular career interests, and become a part of that forum, it's the lurkers who can also become your allies. Headhunters intentionally lurk in business-oriented forums to find people who are qualified for the jobs they're trying to fill, and the way they judge people is by the content and quality of messages they post. Potential employers do the same when they're scouting for management and professional staff. But don't let that intimidate you; just be yourself. It's too casual online to try to be anything you're not anyway.

The deal-making is done behind the scenes, however, in e-mail. Just as it is in the real world, things are also done rather subtly. Unless you've already established a presence and credibility in a forum through your public participation and by developing a network in public discussions and through e-mail, announcing that you're looking for a job doesn't cut it, because they don't know you and have confidence in you yet. Later on, that can work.

But always, it's more a matter of networking. An online forum is nothing more than electronic bulletin board; therefore, even though it's far more interactive than an ordinary bulletin board and much more like a lounge where people gather around the coffeepot or watercooler to talk shop and schmooze, it helps to remember that it's just as much a process as it is a place.

You also have to be resourceful, as Karen Ramos discovered. She's a carpenter and her husband's a welder, and neither was faring well in western New York because of the economy there. She bought a computer to keep her books and organize her business, then one day a disk arrived in the mail from America Online. So she plugged it in, logged on, and started looking for whatever she could find that might help her decide how to improve her family's business prospects.

"You really do have to look around," she says. "I had to find a starting point, so I used a lot of keywords. Some of them work, some don't. Sometimes you have to think of another word that's similar. But they'll get you in the general area, then you can take it from there if you have any common sense."

For her, the first keyword was *small business*. That led her to the Small Business Center, where she found a section titled Women in Business. Then by exploring the Small Business Resources Library, she came across a file called "Help for Women and Minorities." By amazing grace, one of the agencies listed turned out to be an agency in the Raleigh area that helps women in the construction business. Bingo! But Karen had a husband and four children who would be uprooted by this move too, so she needed to know much more information to make a decision.

So it was back to the Go To menu on America Online's main screen, and more keywords. In this way, she found the Real Estate Online area and met helpful agents who sent her packages of information from different cities she was considering. Other tests of her newfound science of keyword searches took her to Compton's Encyclopedia, where she typed in North Carolina and looked up the demographics and data about the economy of several cities. Then she tried the Personal Finance Forum, where Hoover's Company Profiles caught her eye. That triggered another idea. She knew that IBM, Saturn, GE, and several other major companies had moved their headquarters or sizable parts of their operations from New York and other cities in the Northeast, and that they had left for the same reasons she wanted to: high taxes and low profits. So she tracked a few of them through the Geographic Listings, and found a cluster of companies resettled in the Greensboro and Raleigh areas. "That told me something," she says, "so I followed their trail. That means those areas are going to be busy, and those companies are going to lend a boost to the economy."

Well, Karen and her family did end up moving to the Raleigh area, and with more local legwork she and her husband quickly started getting work. Even though both of them had been in favor of the move and she'd certainly done her homework, she had second thoughts at the last moment. It was the people she'd met online who pulled her through that phase, she says.

> *There's an equality there, and we're all in business and have the same goal. There's no discrimination online that I have found, but you can find almost anything else online.*

"I was really torn when it came right down to going. My kids are school-age, and it was a big deal for them to leave their friends. But the people I correspond with are good buds, and are in related fields of work. One's a builder in Texas, one's an investment banker in Denver, so his advice was good to listen to, and there are a few others. They are all men and, I must say, good advisers; they only wanted the best for me, and that's more than I can say for some 'real people' that I know! They gave me a lot of moral support. They'd say, 'Think back to why you wanted to do this in the first place. You're making this decision for a reason.' It was their support and encouragement that gave me the backbone to go ahead with it.

"I don't know them personally, so we're more like business friends. I don't know what they look like, and they don't know what I look like. There's an equality there, and we're all in business and have the same goal. There's no discrimination online that I have found, but you can find almost anything else online. Now you can e-mail all kinds of agencies, because they're all online. You can access the classified sections of some newspapers for jobs in their areas. I can access *Commerce Business Daily* too. For a woman like me who owns a business, that's great. There's all kinds of government work here, and that's how I find out about the contracts. Being able to search it and some of the other publications online just for what I'm interested in saves me a couple of hundred dollars a year in subscription fees."

– 213 –

In case you haven't noticed already, there are three themes in this chapter that are also threads that run throughout the book. To make sure that nobody misses them, here they are:

➤ First, as Karen says, you can find virtually any kind of information online, but you need to learn to think creatively, be resourceful, and look beyond obvious things such as menus and messages to get the most benefit.

➤ The second point is that information is just information, and we're all already overloaded with that already, so the important thing is how you use it and how that makes a positive difference in your life.

➤ Third, the relationships you form online are every bit as important as any information you'll ever find there. Which is as it should be, so all's well and right with the online world, or at least a lot more so than most of the information highway hype would lead you to believe.

# Highlights: Online career resources

The following are online career sources I highly recommend:

## CompuServe

This one has no peer for career purposes other than the Internet special-focus mailing lists. It's better than many Internet newsgroups because, as they say online, the signal-to-noise ratio in too many of the newsgroups sometimes prevents the occasional signals getting through the constant noise. CServe or CIS, as the online shorthand goes, has long been considered the most business-oriented of all the commercial networks and still is, even though they seem to want to shift a bit more toward recreational users.

From a career perspective, CompuServe's strength is in its forums that focus on various professions, each of which has 23 subsections. To find any of those listed here, use the FIND command and just type the name of the forum. Among them are Aviation, Broadcast Professionals, Computer Consultants, Desktop Publishing, Entrepreneurs, Graphic Design, Journalism, LitForum (mainly for fiction writers and poets) Medsig, PR & Marketing, ProPhoto,

SafetyNet (law enforcement and security pros, firefighters) and Working from Home. Several of these forums are home to the most prestigious professional associations in the related fields, although those sections are open only to members, therefore often out of public view. Many of those involved in some way in the computer industry also hang out in the Executives forum in ZiffNet, which you can reach through a gateway (GO exec, but to return to the main service you'll need to type GO CIS if you use CIM, CompuServe's own software, or click on the door icon). The E-Span job database is on CompuServe, and because of the reputation of this network in attracting businesspeople, jobs listed tend to be management and above.

## The Internet and World Wide Web

The very best resource for scouting jobs, moving up in your present one, and all-around career coaching are the Internet mailing lists. They stay very focused, rather than getting raided by rabble rousers or devolving into rambling discussions as the newsgroups do, and the people who take the time to read them daily are savvy and are there for the same reasons you are: to extend their networks and learn. Try a couple about your field until you find the right fit for you. If you make yourself known and keep your comments pithy and concise, you'll establish credibility with others on the list, and they'll eventually tell you about private mailing lists that are open by invitation only to those whom the members and list owner feel would be valuable additions to the group. Lest you bristle at the thought, doesn't the real world work like that? Yes, it does. And this can get you into an inner circle that you might never have had access to otherwise.

There are numerous job-listing databases on the Net, as well as numerous newsgroups related to almost every legal profession (and some otherwise, perhaps). And they're usually searchable, so it beats combing through the Sunday classifieds if you find one that matches your interests or locale.

Therein lies the problem. You could search constantly and never be able to keep up. Thankfully, Bernard Hodes Advertising, a recruitment advertising agency, already thought of that and created Career Mosaic, shown in Fig. 13-1. It searches all those newsgroup ads for you, by keyword, and gives you a customized list. You can also search Career Mosaic for information on companies and university job openings, and browse their library for tips. Make that your first stop. To get there on the Web: http://www.careermosaic.com/.

The other site not to miss is the Online Career Center (see Fig. 13-2), where you cannot only search the job-listings database, but also upload your resumé

**13-1** *Career Mosaic on the Web.*

**13-2** *The Internet's Online Career Center.*

for employers to find when they search for prospective employees. It also leads you to Help Wanted USA, the same service that's on America Online. The Web address is: http://www.iquest.net/occ/ —> Job Junction. Another thing to check out on the Net is E-Span, the same service that's on CompuServe, also on the Web at http://www.espan.com/. Also try the Federal Jobs Mailing List. To subscribe, send e-mail to: listserv@dartcms1.dartmouth.edu. and in the body of the message, put AFD ADD fedjobs package F=mail. And to help your favorite college student decide what to do with the rest of her life (or yourself, because anyone could benefit from reading this), peruse *Princeton Review's* guide on "How to Survive without Your Parents' Money." It's on the Web at: http://www.review.com/career/80102.html.

And when you're done, delete the part after "career" and see what else is there, because that takes you back one level at the same site. That works for all Web sites too, as long as you don't cut off too much. You can't break it, so experiment.

The best career resources the commercial networks have is their browsers for the World Wide Web and access to Internet newsgroups and mailing lists. That said, the commercial nets have a stronger sense of community than any public place on the Internet, so there's still value in making them part of your career network if you find a forum that's a good match for you. Here are some highlights of what's elsewhere:

## America Online

The Exchange section is home to the Careers Board, which has more than 40 subsections for people in different professions. All of these sections were started by members, so if there's not one for what you do, you can start a new folder, as AOL calls their subsections. You'll find some here that aren't on other commercial services, such as a discussion area for property management. Because members can start sections, and they either fly or don't, new ones appear and others languish or disappear. There's also a Small Business Center and sections for people in various professions hidden away within broader-topic forums. Again, with America Online you often have to explore to find things, because in an attempt to make it easy they made it too simplistic and their keyword search feature has limited capabilities. There's also a database on AOL called Help Wanted USA, which is available on the Internet as well. They get their mostly white-collar listings from newspaper classifieds and career consultants around the U.S.

# Delphi

Delphi's draw has been its low rates, full Internet access, and easy-to-use links to browse Internet newsgroups from within related Delphi forums. Thus the main-area, officially sponsored forums have remained sparsely populated and computer-oriented, because too many people were just passing through. Now that they'll be offering Web access with Netscape, the most popular Web browser, and reinventing their whole presentation, it's likely to change. The best bet here are the *custom forums*. They're rather off to the side, so you have to look for them, but you can download a list. All of these were launched at the initiative of members, although there aren't many career-related ones. They do have things you won't find elsewhere, however, except perhaps on the Net. Among your career-oriented custom forum choices: Public Health SIG (SIG means special-interest group), Inventors and Inventions, Nurses Station and the Nursing Network Forum, The Teachers' Lounge, The Job Complex, Work at Home, The Mommy Track, Self-Employment, and Advertising/Marketing Professionals. You might want to check Delphi's Classifieds too, just in case.

# Prodigy

Prodigy's Careers Bulletin Board has sections on management, engineering, medicine, entrepreneurship, and more, and you'll find occasional job-notice sections on business-related bulletin boards (Prodigy's name for forums). The Homelife Bulletin Board has sections for Working, Stay-at-Home Parents and Daycare, and that's a career issue. Like America Online, you have to browse around here to see what's where. But like other services, Prodigy is introducing significantly revamped software, which won't be available when this was written, so you're on your own. If you ask on the Careers BB, though, some kind soul will be sure to guide you. Like everyone else, Prodigy also has a classifieds section, but skip this one.

None of the following rank on the first-call list for career prospects, although some do for other purposes. But here's the rundown:

# eWorld

If you take it as a given that Macintosh programmers and consultants are easy to find on a service run by the people who make Macintosh computers, which they are, and skip the tech zones, then you'll find more musicians than anyone else here, it seems. That's because many of them use Macs to create electronic music. But because theater and dance involve music, you'll also find a fair number of those folks here too, and a lot of graphic designers because they, too, favor Macs. Since many schools use Macs rather than PCs, eWorld is always likely to have more students. eWorld is still so small compared with the services it eventually hopes to compete with, that if your goal is to vastly expand your network so you can connect with those dozen or so people who might become mentors, this isn't the best place to look.

However, one of my favorite places on all the services, called An Income of Her Own, is here; you'll find it in the Learning Center (see Fig. 13-3). It's designed to guide budding entrepreneurs, mainly high-school-age girls, in starting and running their own real businesses. Alas, this section, like almost everything on

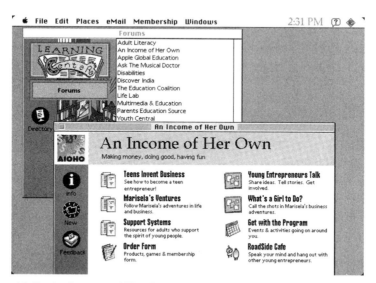

**13-3** *An Income of Her Own, in eWorld's Learning Center, is an online mentoring program for real business operated by teenage girls.*

eWorld, is sponsored, and perhaps for-profit. So not only are the businesspeople who favor DOS computers and who could enrich the experience for the girls not here, there aren't many girls either. Plus, the conversation seems all one-way, meaning the sysop posts messages, but there were no more than about a dozen replies or member-to-member messages a month during the time I monitored eWorld.

## GEnie

GEnie has good classifieds, as online classifieds go, and they're divided into categories. But your best bets here are to roam the Career & Professional Services section and, if it applies to you or might, the Home Office-Small Business Roundtable. Some of the discussions on GEnie are solid enough, such as the Digital Publishing Forum about electronic publishing (MOVE digipub), and it has a loyal lot of members. But unless they regain lost ground as a result of their new software, you'll get more from other services for other professions.

## Regional and specialized networks

Although Echo and the WELL have members from far outside their respective headquarters, each still has a largely regional flavor and membership base. Don't look there for job listings or obvious options, but do get connected if you live in that area, because the people you meet online will be your best resource. Echo is in New York City and the WELL is in the San Francisco Bay area.

## Women's Wire

As mentioned earlier, Women's Wire does have a section called Career Coach, and picks up some of the Net newsgroup job listings. It also has a good section, Women in Multimedia. But it's still relatively regional—although that could change radically this year if their plans pan out—so don't put this on the first-cut list for career help yet. However, one of Wire's pluses is that it's the norm to network, so more goes on behind the scenes.

# Things to come

Interchange Online Network and Microsoft Network were both still under construction when this was written. Both look promising, but it's too soon to say, and will be for a year or so even after they've launched publicly. See chapter 3 for general information and a few specifics. Also, the *New York Times* classifieds should be on the Net by the time you read this.

# CHAPTER 14

# Business management and marketing

In some ways, getting online management advice is as tricky and elusive as finding a business mentor, because it's more about cultivating relationships and learning what information is available online and how to get it than about finding someplace with the sign "Management Advice, Ask Here." The only place that really comes close to the latter is the Management Tips section in GEnie's Careers/Workplace Roundtable. Normally, the best route is to find a newsgroup or mailing list on the Internet or a forum on one of the commercial services that's specifically for people in your profession. Quite a few of each of those exist, but the quality and level of activity vary. In this chapter, we'll look at some of the ways women use online services in their work, but before the "how" let's hear what a few women say about the "why."

# Benefits beyond information

Sure, you can get guidelines online for implementing a smoke-free office policy or writing a personnel manual, or even find someone to hire for a project or job, or find a job yourself if you're even slightly persistent and resourceful. But if you log on only when you need something specific, or lurk around reading other people's messages, or look only for facts and articles, you'll miss most of the less obvious benefits of being online.

## Earn a promotion

Jean Rowan, Director of Marketing for Carneiro, Chumney & Company, a San Antonio CPA firm, mastered so many online research skills that she created a new niche for herself both in her company and in her community: "I'm now on the business evaluation team [of the firm] as an economic researcher, which is very exciting," she says. Jean also makes presentations about using online services, does online research for clients in her spare time, and has begun writing occasional articles for the local paper. In fact, the reporters and editors sometimes call her for help on how to find information online. "I've developed a good enough friendship that I don't have any difficulty getting something about the company published," she says.

"Too many women think because [being online] is computer-related, it must be technical. It's people-related. The computer is just a tool to reach out and touch so many more people and other women in your profession."

## Increase your confidence

Barbara Bennett, owner of Sarasota Occupational Therapists, in Sarasota, Florida: "Participating in discussions in the Medsig Forum has increased my confidence, especially in talking with doctors and people I associate with, and that's spilled over to how I deal with people locally, offline." Many other women report that participating in heated issue-focused discussions with men online has helped them learn how to get their points across without withdrawing, placating, or getting angry. They've learned to assert themselves in ways they weren't previously accustomed to doing, but now do comfortably both online and offline, thanks to the opportunity to try out new behaviors, gauge reactions in a relatively nonthreatening environment, and gain confidence in themselves as a result of the process.

*Too many women think because [being online] is computer-related, it must be technical. It's people-related. The computer is just a tool to reach out and touch so many more people and other women in your profession.*

## Extend your network

Lila Roebuck Canuelle, Director of U.S. Development for Pro-Connect: "I've made some long-term connections with individuals online whom I trust and now feel comfortable calling for any kind of help with a personnel issue. I could not recommend a more effective network."

## Break down barriers

Marcia Layton, business plan consultant, Rochester, New York: "I've found that I'm probably taken more seriously online than in some of my dealings with investors face to face, primarily because my youthful appearance has made some potential clients somewhat apprehensive. While I can overcome that in time, it lengthens the selling cycle. Online, no one can see that I'm closer to 30 than 50, and they take my advice at face value rather than discounting it because of my age."

# Gain flexibility and independence

Sarah Browne, content architect, Milwaukee, Wisconsin: "Master the mighty modem, and you have bargaining power to negotiate with your employer, to telecommute or work from home part of the week, or gradually segue into running your own home-based business." Sarah spent several years in Manhattan advertising agencies before fleeing to a lakeside home in Milwaukee, Wisconsin, where she built a practice as a marketing consultant and now advises companies on how to reach women online. "All I know is that every afternoon when my Katie comes bouncing off the school bus I'm grateful for the gift of technology," she says. "Without it, I'd still be in some high-rise box in Manhattan."

> *My husband at the time told me I'd probably be a bag lady*
> *if I tried self-employment. Four years and one divorce later I*
> *was making $75,000 a year.*

Shelly Espinosa, corporate trainer, Denver, Colorado: "Eight years ago I was making $7.50 an hour working as a secretary. Most of the books and seminars for secretaries were produced by people who'd never done the job, but knew how we could do ours better. I had an audacity attack, got a couple of friends together, and wrote a book called *Working Solutions from Working Secretaries*, which offers field-tested ideas from support staff around the world. I used the book to build a business called Working Solutions, which specializes in seminars for support staff. My husband at the time told me I'd probably be a bag lady if I tried self-employment. Four years and one divorce later I was making $75,000 a year." Much of what she learned about how to build her business she learned online, as she explains in the last section of this chapter.

## Make yourself more marketable

Lorraine Sileo, editorial director of the Electronic Services Group of SIMBA, a media industry analyst and publisher in Wilton, Connecticut: "You have to keep up on a daily basis to look smart now. I think women are great at networking, and they'll find other ways to do it, other than online. But it's the knowledge and the intelligence you gain that still makes it important to be online.

"Women are so used to doing things for other people or making their bosses or other people happy. They think they have to have something to show how they spent their time at work, like a brochure they produced or something like that. They don't see that what they learn online makes them more interesting, better-rounded, better-speaking, more intelligent people. It also opens up options outside of your current work position. So you do it for yourself, because it makes you smarter and leads to other opportunities. Men are always looking over the fence more than women."

– 227 –

## The online wiles and ways of corporate managers

 When Susan Corcoran of Family Investment Trusts in St. Louis asked for information in America Online's Small Business Forum about writing job descriptions for staff positions for a nonprofit agency, she was disappointed in the responses she got at first. The early responses were pretty elementary, because she hadn't made it clear that she already had some expertise in this area and was looking for specifics. Her original message read:

```
Does anyone have any sample job descriptions that I could use to get
started in writing job descriptions for administrative secretary,
communications director, and executive director of a small nonprofit
agency? Also, sample personnel policies. Or are there some to download?
I sure didn't see any. Please e-mail. CorcoranS. Thanks.
```

Later Susan decided to give it another chance, so she checked back to clarify and see if there were more responses, and was quite pleased with the results.

"We were writing a personnel manual and developing policies," she says, "and we got good advice about what to be sure to consider and an outline of a manual to start from."

 Valerie Hart, VP of Marketing for Celutel, a cellular telephone provider based in Annapolis, says they use Dow Jones News Retrieval, Prodigy, and CompuServe on their LAN to monitor stocks, industry groups, and trade publications, and to make airline reservations. "As a company, you have to know what's going on to be competitive," she says. "Personally, you've got to maintain your own competitiveness too if you want to provide some direction and be active in your career."

*I was spending a lot of a time at the library, then I found out that I could do 99 percent of what I was doing from my computer rather than going to the library.*

Lila Roebuck Canuelle, of Pro-Connect, a Mattawan, Michigan-based company that provides employment advertising for companies nationwide, particularly in engineering and technical fields: "When we decided to make our offices smoke-free, we had several employees who were smokers who already felt they were being ostracized, so I really wanted to talk to people who had already integrated a smoke-free policy into their workplace, and find out how they had made it go smoothly and how they had taken feelings into consideration." A human resources director at another company uploaded a copy of their policy to her, and gave her suggestions based on his own experience. "He advised me to make the creation of the policy a team effort—to send around a questionnaire to find out how people were feeling, and then talk to them," she says. "I followed his advice, and was able to avoid a lot of mistakes."

Jean Rowan, marketing director for the CPA firm mentioned at the beginning of this chapter, uses CompuServe for its magazine and industry newsletter databases and Delphi for its easy and extensive Internet access. "As the director of marketing, I do all of the industry research. I was spending a lot of a time at the library, then I found out that I could do 99 percent of what I was doing from my computer rather than going to the library."

"Using online services has helped us because we have a lot more to offer clients. For instance, when we were getting ready to do a report for a client who is a food processor, we looked at their financial ratios and saw that their inventory was higher than it should have been. So I went online to get comparative information about other food processors, and to find out what they did to reduce their inventory and use that space for something else. I pulled down five or six articles from industry magazines and newsletters, which I passed on to the CPA on that account, which he passed on to the client. That was really helpful for me, because there are a number of different smaller publications online that only people in the industry would be reading.

"If we have a client that's looking for industry-specific software, I'll use the Accounting Software Vendors Forum or a forum about that industry. I found a cost-accounting program specifically for a commercial bakery that way, and found a local users group for that software our client was able to talk to people who knew about it."

# Why entrepreneurs have the edge online

Online services serve as constant consultants for solo businessowners of very small businesses, because they can learn from colleagues any day and every day, even if they don't have co-workers. From tradename research and writing a business plan to expanding internationally, there's help online every step of the way. Many of them find clients or customers online too, or cultivate the ones they have by keeping in touch by e-mail. The discussion groups are almost as important as all of that, however, because they substitute for the shoptalk that those in corporate settings take for granted.

"If you're an entrepreneur today, you're very decentralized," says Mary Anne Graf, who runs Health Care Innovations, a million-dollar company in Salt Lake City, and its sister company, HCI Market Research Group. "What you really miss is the corporate watercooler, and that's what online services often are for me."

*There are some real authorities on the forums—people running multimillion-dollar companies who are willing to share what they know freely. The depth and quality of the discussions amaze me.*

 Mary Anne's core staff is small, but she has a pool of 30-some consultants across the country whom she pulls in for different projects, a trend that's common to both big corporations and small businesses. Therefore, she does much of her business with clients and contractors by e-mail, which enables her to work from home when she needs to, even if just for a morning. She relies on various forums for advice, depending on the situation. For instance, before she went on a speaking tour in Japan, she asked people in CompuServe's travel forum how to dress, and found out that her favorite vibrant colors would not be the best wardrobe for a woman to be well-received in that culture. In the PR and Marketing Forum, she learned how to conduct telephone focus groups and hired a PR consultant she'd met there.

– 230 –

"There are some real authorities on the forums—people who are running multimillion-dollar companies who are willing to share what they know, freely," says Mary Anne. "The depth and quality of the discussions amaze me."

Barbara Burnes, a strategic planning consultant based in Westport, Connecticut, says being online has saved her more than once. She uses online databases to research everything from esoteric information on welding masks to trends in the wine industry in Italy. When a crucial business-simulation group she'd put together fell apart, Burnes posted a call for help on CompuServe. It was a long shot, but it was that or risk losing her client.

"The whole thing could have blown apart on me," she says. "I posted messages in about five different forums, and got messages back from all over the place." That helped her get a new group together again within a couple of days. "This was not a simple thing. They had to have the right kind of business background. There's no way I could have put together another group that fast with qualified people otherwise. I was astonished!"

Alice Massoglia joined GEnie for other reasons, and says she didn't expect it to benefit her business, but has been pleasantly surprised. "My husband and I have a used bookstore, and also sell used and collectible books at various science fiction and collectors' conventions," she explains. "I use the Science Fiction Roundtable to keep track of convention news and updates, to get information about authors and their upcoming books, and correspond with readers, fans, friends, and customers via e-mail. I've also downloaded graphics programs which we've used in generating flyers, signs, and newsletters, databases for organizing request and mailing lists, and a program for converting an existing set of lists from one computer to another. I'm currently trying out a bookkeeping program that I chose on the basis of discussion on the Home Office/Small Business board, and a time management/organizer program that my friend downloaded recently. I think I've gotten an incredible amount of information and valuable downloads."

 Gerre Witte, owner of The W Group, a design and copywriting agency in Grand Rapids, Michigan: "Once I needed to research color trends for a client that manufactures hospital equipment, and wanted to know what accent colors they should paint a new cart. I found some pertinent articles and information through the magazine databases and got copies of the articles, which then led me to a national color forecasting group. I called them, and as a result of the research we were able to make authoritative recommendations to the client, and turned the whole job around in about two weeks."

*There's no way I could have put together another group that fast with qualified people otherwise. I was astonished!*

Remember Shelly Espinosa, the author of *Working Solutions from Working Secretaries*? She says maintaining a presence online for herself and her training company, Working Solutions, helps her build name-recognition and credibility, as well as guides the development of new services and content of her programs.

 "Because of the online services I'm able to tailor any seminar topic to a specific work group," she explains. "I teach a class called Managing a Maximum Workload, and when I got a call from an engineering company to teach that class to their engineers, I went online for advice from engineers. I got ideas from them as to how they actually deal with too much to do in too little time."

Shelly's online home base is the Trainers and Human Resources section of the Working from Home Forum on CompuServe, but she also monitors other forums to get ideas and spot trends in time to take advantage of them. "I design all my own seminars, videos, and audio tapes, so I listen to what people gripe about," she says. That's what gave her the idea for the Managing a Maximum Workload seminar, in fact. "I can also problem-solve with other trainers and consultants on dealing with clients and determining client needs. . . . And the support of other single parents has been wonderful."

# Highlights: Online management resources

## For entrepreneurs

*Major commercial services* The Home Office-Small Business Roundtable (HOSB) and Careers/Workplace Roundtable on GEnie, and the Small Business Center on America Online are all run by Janet Attard, a communications consultant and author of *The Home Office and Small Business Answer Book* (Henry Holt, 1993). Paul and Sarah Edwards, authors of several home-based business books that have become standards on the topic, run the Working from Home Forum on CompuServe, which takes its name from one of their books. All of these, as well as the Business Bulletin Board on Prodigy, are excellent resources for home-based and very small businesses. *Home Office Computing* magazine maintains a section on America Online, and offers pay-per-view articles on Prodigy, and *Entrepreneur* magazine sponsors the Entrepreneurs Forum on CompuServe, which has more of an international focus.

*Entrepreneurial Edge Online* This is an online tutorial service operated by Telebase Systems' Corporate EasyNet, with more than 100 modules to teach you such things as how to analyze your business finances, expand through new product development, identify a target market, prepare a customer profile, create an effective ad, hire a sales staff, write a sales proposal, buy a business, develop and use a business plan, prepare a market analysis, or establish the right marketing mix. The annual subscription is $49.95, including unlimited time online, but there's a $7.50 charge for each tutorial you choose. For information, call 800-220-9553 or 610-293-4700 in the U.S.

***Entrepreneurs Online*** According to Pam Terry, Vice President of Product Development, Entrepreneurs Online is designed to help people start, buy, or grow a business. It's mainly for businesses with 20 or fewer employees or less than $20 million in revenue. Although it also has forums, some of its databases could be worth the price of admission alone. Among those are Equitied Employment, for people who want jobs but want a stake in the business, or equity, as part of their compensation; Strategic Alliances, for those who want to match what they have with what another company has that can help both businesses grow, such as a specific customer base or a certain technology; Emerging Companies, which offers public companies an opportunity to list their corporate profiles online so they have a better chance of reaching potential investors; and Venture-Net, which allows businesses seeking capital to upload executive summaries from their business plans and information about how much money they're looking for and for what purpose. Entrepreneurs Online is accessible directly or via the Internet once you're a member. The starter kit software is $29.95 and is available for Macintoshes, DOS, and Windows. There's a monthly fee of $14.95, plus nine cents a minute after you've used your basic allotment of four hours each month. For more information, call 800-784-8822 or 713-784-8822 in the U.S.

***The Internet and the World Wide Web*** The main newsgroup is called misc.entrepreneurs, but if you want to avoid the get-rich-quick messages, join the moderated version, called misc.entrepreneurs.moderated. The Small Business Administration and many other resources are on the Web, but the mailing lists are most useful for specific professions. You'll have to search for those by keyword, but you can often get leads from people in the forums on the commercial services.

***The Meta Network*** Run by Metasystems Design Group, a management consulting organization in Arlington, Virginia, The Meta Network (see Fig. 14-1) provides full Internet access as well as unusual services for small businesses, boards of directors, corporate project groups, and government task forces. The concept of their service is based on support for group decision-making, plan-

The Meta Network

Welcome to **The Meta Network Home Page and WWW Server!**

Summary Information about this TMN server can be found here.

Here's our MetaNet Frequently Asked Questions list.

**The Meta Network** is owned and operated by Metasystems Design Group, Inc. (MDG).

**14-1** *The Meta Network's informational site on the Web.*

ning, and project management, as well as consensus-building in organizational development. TMN will set up private areas to enable groups to work together from different locations, and even help facilitate the process. Some companies also set up systems to disseminate information to the public through TMN via e-mail lists and FTP or Web sites.

Their brochure reads: "Metasystems Design Group's business is designing and implementing virtual organizations. We focus on strategies that enable an organization to reflect on its overall quality in the framework of its larger social purpose, and to learn from its experience. We work with organizations to design teams and strategies for change involving patterns of values, language, concepts, people, relationships, media, technology, new structures, and new stories . . . . In particular, we promote cultures that value curiosity, candor, cooperation, and creativity." Lisa Kimball, one of the owners of The Meta Network, is one of the sharpest and most articulate women around, particularly about the capabilities of online communication, so anything she's in charge of is worth investigating further. For more information, send e-mail to info@tmn.com, visit http://www.tmn.com on the Web, or call 703-243-6622 in the U.S.

## For corporate managers

*Commercial services* Check the Careers/Workplace Roundtable on GEnie, which has a section for women, as well as the forums listed in chapter 13, *Career guidance*, for leads to resources for specific professions.

***The Internet and the Web*** Here too, the Internet mailing lists devoted to highly specialized topics and specific professions are the best bet. If you're dreaming of working from home even part of the time, you'll get good ideas for bolstering your case with the boss if you join the Flexible Work Environment mailing list by sending e-mail to listserv@psuhmc.hmc.psu.edu with the message SUBSCRIBE flexwork *your e-mail address.* Here's a sampling of available mailing lists, particularly those for women in jobs where more of their counterparts are still usually men than women:

➤ EduCom-W, moderated list for educators about using technology in education (listserv@bitnic.educom.edu)

➤ FemeCon-L, for female economists (listserv@bucknell.edu)

➤ FemJur, Feminist Juriprudence listserv (listserv@suvm)

➤ GeogFem, for female geographers (listserv@ukcc.uky.edu)

➤ Systers, a private list for women in computer science or computer engineering (systers-request@pa.dec.com)

➤ SWIP-L, for members of the Society for Women in Philosophy and others (listserv@cfrvm.cfr.usf.edu)

➤ WiPhys, moderated list for women in physics (listserv@nysernet.org)

➤ WiseNet, for women in science, mathematics, and engineering (listserv@uicvm.uic.edu).

➤ WITI, Women in Technology, Inc. (on the Web at hHp://www. witi. com or e-mail to info@witi com)

# CHAPTER 15

# Online marketing methods

There's a rather strange phenomenon that affects many normally savvy businesspeople the first few weeks or months they're online. Perhaps some random electrical discharge when the modem connects makes their synapses misfire, because they suddenly forget everything they knew about effective target marketing. The next thing you know they have a team of multimedia developers working on creating a Web site they can call their very own. Then they start casually mentioning "our Web site" or "our home page" at luncheons and to strangers in the checkout lane at the supermarket. Unless their listeners are online and also afflicted by this syndrome, hardly anyone ever bothers to ask "Your *what?*" These born-again Net marketers are usually oblivious to the glazed-eye responses their evangelism engenders.

While their Web site is "under construction" or still a castle in the air, some of these would-be Net marketeers master the intricate art of sending group e-mail with whatever software they use. "Eureka!" they cry. "This is infinitely easier and more cost-effective than paying printing and postage for a 20,000-piece direct mail drop." So they promptly promote their product or service by e-mail to hundreds or thousands of totally uninterested people. This is known as "spamming the net." On commercial services, it's against the written rules. On the Internet, it's against the unwritten rules. Either place, it's likely to result in people retaliating either by private e-mail or public newsgroup flames, or by rigging up a system on their computers to send 10 junk-mail messages to the perpetrators every day for the next month.

# You must understand the content and the culture

A far more innocuous but often equally ineffective method of advertising online is the one Caryn Cain tried when she was new online and hadn't yet learned the lay of the land, much less the culture. She owns New Era Direct, Inc., a marketing and desktop publishing company in Osh Kosh, Wisconsin. Upon discovering the online classifieds, she thought she had found a quick, easy, and inexpensive way to sell her report on home-office taxes. Good direct marketer that she is, she was smart enough to test different headlines and body copy first in the CompuServe classifieds. "After running various ads, getting responses by e-mail, and snail-mailing an information packet, I got nary an order," she says. Thus, her initial assessment of the value of online classifieds became: "I was able to test my concept cheaply on CompuServe and avoid sinking a ton of money into a larger mailing or space ad. The beauty of online classifieds is that you know right away if there is interest, and how much."

Now Caryn realizes that she hadn't explored the online alternatives enough to find out the best way to reach her real target market: not just anyone who might happen to read the classifieds, but home-based businessowners. Had she done that, on CompuServe alone she would have discovered the Entrepreneur's Forum, sponsored by *Entrepreneur* magazine, and the Working from Home Forum, both of which are for the exact people she wanted to reach.

However, she might also have reconsidered whether her report could measure up to the competition there, because the Working from Home Forum has a Tax Section populated by CPAs, enrolled agents, and tax lawyers who specialize in serving small businesses. Not only do they offer free advice by participating in the forum discussions, some of them sell special reports that are in direct competition with Caryn's, plus they have the credibility of the CPA or EA after their names.

With a little more exploration, she would have discovered that nearly all the online services have forums for self-employed people because so many of them rely on computers and online services to run their businesses. The forums are, for them, a combination of get-togethers with co-workers, professional associations, and water coolers where they stop by to say "How's biz?" or learn how to manage their businesses. Indeed, they might be interested in just the kind of tax report Caryn wanted to sell. So would the people who participate in several of the newsgroups for entrepreneurs on the Internet, which are the equivalent of forums on the commercial services.

But how can you sell anything to any of them when it's against the rules? Ah, I'm glad you asked, because when you can answer that question you'll understand the real strategy to online marketing.

"It takes some real know-how to know which newsgroups to post to and how to properly construct your post so it's not offensive," says Jim Sterne, who produces the Marketing on the Internet Seminar in several U.S. cities. "Then, because this is an interactive medium, you have to make it not something that people read, but something they do. . . . You also have to make sure that you're hitting newsgroups that are on a topic related to what you're selling. If you want to sell bicycle tires, it's okay to go into a group that's discussing bicycles and say, 'Oh by the way, we have some information about bike tires over in this other place. Come take a look.' But it's not okay to post that in the scuba diving newsgroup or to post something that says, 'We've got great bicycle tires, they're on sale, call us today.'"

# The stampede to the World Wide Web

By "this other place" Sterne means "a Web site." Eugenie McGuire, who set up her own Web site and designs them for others, says, "I think everybody who's anybody will eventually have their own Web site." Given how anxious everyone seems to jump on the bandwagon, she might be right. But that doesn't mean it's the right thing for your business or all businesses to do, or within the capabilities of your budget, or even that it's the best way to present your product or service at all. The whys and wherefores of the decision are beyond the scope of this book, but it comes down to the same decision of whether to use classifieds or direct mail or any other marketing method, which means asking yourself the same questions you would in the real world: Who is your audience, and what's the best way to reach them? Think it through.

Jim Kinsella, Managing Editor of Time Warner's Pathfinder service on the Web, says there are three things you need to consider in evaluating whether a Web site is a workable medium for your message:

➤ "Evaluate your assets. What's already digitized [in computer files]?

➤ "Is there a metaphor to describe what your company does that suggests a design for your home page?

➤ "Are you willing to make a commitment to using this Web site in the most effective way, which is as a communication device? Do you have a way for people to correspond with you? This is an organic thing, so you must institute a staff, and that staff should be responsible for overseeing the Web site. There must be one person who is the Web master, and who is full-time."

Says Rosalind Resnick, co-author of *The Internet Business Guide* and publisher of Interactive Marketing Alert, an online newsletter: "The Web is definitely the big opportunity for the future. It's going to be an enormous opportunity for business. . . .

*The Internet is a much better place*
*for advertising than for actually selling.*

"The best Web sites, like *Hotwired*, have eye-catching graphics or even video and sound, but they also have a strong interactive component. You can click on the Web page on your screen and send e-mail directly to them. In some cases, clicking on one button on the screen with your mouse takes you to their bulletin board or forum."

– 241 –

But even though Rosalind is in the business of advocating marketing on the Internet, she concedes, "The Internet is a much better place for advertising than for actually selling."

If you're undaunted, the first thing you need to do is make arrangements with your Internet SLIP/PPP access provider (see appendix B) to store your Web pages on their server, or computer. There's typically a one-time set-up charge, then a monthly maintenance fee. For what are known as *vanity pages* or *personal pages* for individuals or families, you can pay as little as zilch for set-up and $10 a month. But for businesses, set-up charges range from $75 to $5,000, and the monthly fees can range from about $150 to $500. Be sure to research the reputation of the provider, consider rates and exactly what you get, and make sure you're comparing apples with apples, not oranges.

The next step is to register your domain name with Internic. Yes, Inter*nic*, not Inter*net*. To find out about that, use gopher to get to:

gopher://rs.internic.net:70/00/rs/ftp/templates/domain-template.txt

When you're ready to start designing and building your Web pages, there are hundreds of people now billing themselves as multimedia developers. Predictably, some are better and more experienced than others. Rates range from about $150 to $3,000, depending on whom you hire and the complexity of the job. Look at their previous work for others, get references, and talk to those people if you choose to hire a consultant. There are many software tools now available to make the job fairly simple to do yourself if you can forego razzle-dazzle features and design. If you decide to go that route or just want to find out what's involved, CyberWeb Software's Web Developer's Virtual Library at http://www.stars/com is a good place to start (see Fig. 15-1). The following Web locations also have tutorials and even software tools and graphics you can download and use:

http://www.ncsa.uiuc.edu/general/internet/www/htmlprimer.html
http://www.willamette.edu/html-composition/strict-html.html
http://www.ziff.com/~eamonn/crash_course.html
http://akebono.stanford.edu/yahoo/computers/world_wide_web/html/
http://www.indiana.edu/ip/ip_support/learn_html.html
http://www.uwtc.washington.edu/computing/www/doswindows.html
http://scholar2.lib.vt.edu/handbook/handbook.html
http://web66.coled.umn.edu/cookbook/contents.html
http://www.rip.edu/Internet/guides/decembj/itoools/nir_systems/www.html

If you can't get to these sites because you aren't on the Web yet, stop! Right now! Don't even think about creating a Web site or doing any online promotion until you *are* online and have been long enough and far afield enough to truly understand what's there and what works, and whether you should even be considering it at all. Yes, people have commissioned others to create Web sites for them without ever exploring the Web themselves, which is about as smart as sending money to someone in Las Vegas to throw at the roulette wheel for you.

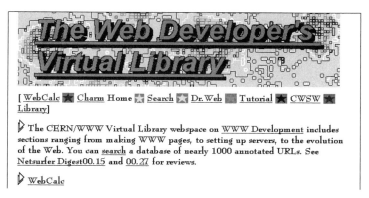

*15-1 CyberWeb's Web Developer's Virtual Library at http://www.start/com. It's a good place for would-be Web site creators to start.*

Finally, the following books come highly recommended by people who have already tramped this terrain:

*HTML Manual of Style*
by Larry Aronson
Ziff Davis Press, 1995
ISBN: 1-56276-300-8

*Teach Yourself Web Publishing with HTML in a Week*
by Laura Lemay
SAMS/Macmillan, 1995
ISBN: 0-672-30667-0

# Getting people to buy by e-mail

Let me be emphatic and clear: It is not cool to send unsolicited e-mail promoting your product or service. It's even against the rules sometimes, and you could be denied access by online service providers if you persist in doing it after they warn you to quit.

"Use the print paradigm," advises Kinsella. "You wouldn't flood your message into every household in America, nor should you send your message to every newsgroup. That's very obnoxious, and not the way to go about it."

People have already found ways around that, of course. *Positive-option mailing list* is a euphemism for "get people to ask you to add them to your mailing list." The way you do that takes some homework, but here's how it works:

# Building e-mail lists from the commercial services

**Step 1** Find out where your prospective clients congregate online. Thus, if I want to find people who might be interested in subscribing to *Freelance Success*, the newsletter I publish online for writers, I would join each of the major commercial services for the trial period and look for forums for writers. Then I would browse through the message threads and libraries there to see whether it looks like the content of the discussions and the library files is similar to the content of my newsletter. If that proves promising, I would then look for the subsections within the forum that are an even closer match. In my case, because my newsletter is for journalists and nonfiction writers, I would skip the sections for poets or fiction writers. Not that they don't overlap, but we're talking targeted marketing here, remember? So I want to test the narrowest, most focused list I can compile.

**Step 2** Explore until you figure out how a particular service lets members list information about themselves. On America Online, for instance, members' brief biographies are listed only in the system-wide directory, whereas CompuServe's system-wide directory lists only name, location, and ID number, and any details about members' professions or interests are in the forum membership directories. In both cases, details are there only if the member chooses to include them, so you won't find all likely prospects that way. Whether it's a master list for the whole network or one within a special-interest forum, however, you can search these directories by keywords that indicate members' interests. Thus, for my newsletter example, I would use *writer*, *freelance*, and *journalist* as keywords. On most networks, you can capture the results of that search and save it to a file on your computer.

***Step 3*** Learn exactly how to edit that file and convert the names to a group e-mail list, and how to send mail to that list with the particular software you're using to connect with that network. In most cases, this is a lot easier than it sounds, even if it is a tad trickier than just posting a classified advertisement. But direct marketing in the real world requires more work and has a longer learning curve than faxing an advertisement into the local newspaper.

# Building e-mail lists from the Internet

The process is similar on the Internet, only this time you're looking for relevant mailing lists (also known as *listservs* or *listserves*) and newsgroups. For mailing lists, find and download the List of Lists or Publicly Accessible Mailing Lists. They're very long, so it will take several hours to cull through them to find only the few lists that fit your criteria, although you might get lucky and find likely lists by using America Online's Internet mailing list search feature. The newsgroups are easier to find through any service with Internet access, but there are thousands, with some added and some dropped daily (see chapter 2, *A guided tour*, for information about categories of newsgroups). There's no definitive, all-inclusive listing of newsgroups yet, and probably never will be because they change so frequently. However, the Yahoo List, the University of Michigan's Subject-Oriented Clearinghouse, and other resources are excellent places to start. (See appendix G.)

Then you need to subscribe to those lists and newsgroups for at least a couple of weeks, read the messages daily, and determine if it's actually a good match. Better yet, after you've lurked a while to get a sense of who's saying what and what you might have to contribute, follow Sterne's advice again and get involved. This also enables you to do indirect promotion through your signature line, as Jim Sterne mentions. But be forewarned: the volume of e-mail you'll get from many of the mailing lists and newsgroups can be overwhelming and very time-consuming to wade through. It's less so on the mailing lists, which are often moderated or edited so they stay more on topic, thus you won't have to spend as much time sorting the wheat from the chaff. As they say on the Net, the signal-to-noise ratio in some of the newsgroups is more noise than signal.

If, and only if, you find a mailing list or newsgroup whose subscribers are likely to be interested in what you offer, there's a way in most cases to send an e-mail message to the main computer that manages that list or newsgroup—the same one you sent the message to when you subscribed—and request the list of everyone who subscribes to that list or newsgroup. This is all handled automatically by the computer, so you don't have to justify your request.

## What to do with the lists once you have them

Here's where the positive-option mailing lists come in. Once you've learned how to send group e-mail or what amounts to a mass mailing with whatever software you use, send a very brief (three- to five-line) introduction of your service or product to people on the list. Be a bit apologetic about invading their private e-mailbox, tell them what you offer and why you thought they would be interested, and give them a way to respond and tell you that they want to remain on your mailing list or receive more information. It probably goes without saying, but be sure to save the e-mail addresses for anyone who ever inquires about your product or service. Eventually you'll have a truly targeted list. But try to keep anything you send thereafter to less than 20 lines, which will irritate the people on your list less and cost you less.

*Warning:* Individuals own the mailing lists. The lists are usually free, and the owners usually derive little or no direct income from all the work they put into maintaining them. If the privilege of automatic access to the list of people who subscribe to the list is increasingly abused, the owners are likely to quit making those lists available. So follow the Golden Rule, please.

## Other kinds of online marketing

If there's any single key to successful online marketing, whether it's on the commercial services or the Internet, it's that public-relations techniques are far more effective than obvious advertising. Particularly on the Internet, people strenuously object to being hit on to buy something, especially via e-mail or in the newsgroups. They object to information being foisted upon them rather than choosing to go after it themselves. Even when they do consent, they want information with a service aspect, not a pure sales pitch.

The other key is that there's no single answer. As always in marketing, it's a mix. "You have to use all these online tools together," says Rosalind Resnick. "Sometimes it's the low-cost, low-tech approach that works just as well or better. Although a message on a forum or newsgroup or mailing list scrolls off so people might not happen to see it, the Web site stays there as your own interactive bulletin board in cyberspace."

 *Tip*: There are media e-mail address lists available on the Net, but don't bother. The addresses are rarely valid because savvy reporters and editors change their addresses as soon as they start getting bombarded by junk e-mail, and press releases need to be sent to the right person. So you still have to do your homework.

There are numerous lists, Web sites, newsletters, and newsgroups that announce new online services, so if you create one, invest the time before you launch your service to find out who compiles these lists and let them know about your service. If you establish a Web site, also let any related Web site operators know about it, because they might create a link from their site to yours. That exposes people to your offer who are at least interested enough to visit your site, without additional effort or expense on your part.

> *You have to use all these online tools together. Sometimes it's the low-cost, low-tech approach that works just as well or better.*

As Sterne says, you also have to understand the online culture. The tradition and the norm is that people freely share information and knowledge, and help one another. This is another instance where the online world works just like the real world. As the sales trainers say, always remember that people buy from other people. So it's vital that you establish a relationship, trust, and credibility, all of which count as much as what you're selling. People do tend to do business with others they meet online, partly because people online still feel they're part of a quasi-elite community, so they feel a kind of loyalty to one another.

Whatever online methods you decide to use to promote your business, always bear in mind Sterne's advice to provide information that's useful, not just hype. "The classic approach is to get involved in the conversation and be helpful," he says. "'Buy my stuff' is not help. . . . In a newsgroup [or forum], if I'm trying to sell something, I'll join the conversation and talk about what they talk about and stay on topic. In my signature, it will say 'Marketing on the Internet Seminar.' It's a business card. It also helps people identify you by a name or a slogan. As long as the subject matter in your post is on-topic, helpful, and giving, you become known as a voice of reason."

The other thing to remember if you want to succeed long-term, after the hoopla about the Net and the Web has long-since died down and it has become just another option in life, is that anything online must have an interactive component, or else people stop by once and ignore it thereafter. Therefore, you need to devise a way for people to respond and take some initiative, but something beyond the traditional direct mail gimmick of a sticker to peel off and move from one corner of the reply form to another or a box to check. That means you have to design something specifically appropriate for this medium, not just dump data and go back to business as usual.

There was great resistance to turning the Internet into an opportunity for people to make money, and we're still in the early stages of that development. So if you go online, go with an eye to the future rather than immediate rewards, although those can come too. You might find that the process is fun and valuable to your business in other ways, even if it's not immediately profitable.

"What I love about the Internet," says Rosalind Resnick, "is that it's like somebody gives you a bunch of blocks, and you can build whatever you want with them."

*Tip*: The following mailing lists on Internet marketing are also worth trying: Free-Market, Market-L, INet-Marketing, HTMarCom, and HTMNews. Also be sure to see the Bibliography for references to a few of the better books on Internet marketing.

# CHAPTER 16

# Managing your money by modem

According to an August 1994 article in *Consumers' Research Magazine*, ". . . U.S. households that already have access to the major online services are being exposed to hundreds of fraudulent and abusive investment schemes, including stock manipulations, pyramid scams, and Ponzi schemes. . . . At least [five] states currently have investigations underway involving suspected investment fraud and abuse in cyberspace."

If anything, it's getting worse, not better, as more legitimate businesspeople try to exploit the opportunities and more con artists try to test the boundaries. To get around the typical rules or sanctions against commercial promotions, people who are actually hyping something because they sell it or have a vested financial interest in it sometimes post ostensible tips as though they're ordinary investors who've just heard hot news. As in the real world, there are also people who simply get their kicks from misleading people and promoting get-rich-quick schemes. However, if you use common sense, there's still a wealth of useful information to help you make investment decisions. The hope that many people had that online access would turn into a way to bypass brokerage fees and traditional methods of research and trading has only partially materialized, and isn't likely to happen completely. What's happening instead is that traditional financial services, such as brokerage houses, are creating their own online networks and selling special software for them. People are beginning to switch to those services, although they sometimes still use the online forums on mainstream services and Internet newsgroups, such as misc.invest.stocks, for discussions.

# What service is best for which purpose

Nearly all the women interviewed for this chapter say they're more wary of investment newsgroups on the Internet than they are of the forums on commercial services, because the Internet newsgroups have no one in charge as the forums do. "The Internet is not a place to have serious discussions about anything," Robyn Green of Miami, who is a former sysop in the Investors Forum on CompuServe, says flatly. "Part of it is because the messages aren't threaded [connected]. It might sound strange that the software has an effect on the nature of the discussion, but it can. The messages on CompuServe can be much longer than on other services, and you can write messages offline and then post them. So people write very thoughtful, two- and three-page messages."

Robyn's point about the disconnected messages is true with most software, but newsgroup messages are threaded in America Online's Internet area, even though they aren't in AOL's own forums.

Ilana Stern follows the messages in the misc.invest.stocks newsgroup and frequents several sites on the Web. Her retort to those who are wary of the Net is, "The intelligent, educated person—hell, just the average common-sense skeptic—can quickly see what's a scam and what's not. I often get ideas from the Net, but I never invest without doing a little of my own research in the library."

Although Ilana uses only the Internet rather than the higher-cost commercial services, more women seem to prefer commercial networks instead for investment guidance, and sometimes use more than one. To the extent that you can get investors to agree on anything, the consensus seems to be that CompuServe still has the most in-depth information and the most carefully moderated forums, but that members of the Investors Forum there, particularly, are so serious and so experienced compared with forums on some of the other services that Prodigy, America Online, or GEnie can be less intimidating to a more casual or novice investor.

The women I interviewed by phone or e-mail consistently commented that because there are more resources on CompuServe and they're often more complex and scattered around various services on the system, it takes more time to learn what's there and how to use it well. Therefore, for one-stop, at-a-glance information, Prodigy is currently an equally popular service. GEnie used to be one of the leading services for personal finance (see Fig. 16-1) and does have excellent resources, but most are surcharged and it lost much of its lead when it didn't keep pace with software developments. In the interim, America Online, which was considered rather weak in this area only a year ago, has continually added services and is now in third place in popularity for personal finance, particularly mutual funds. But the lack of threaded messages in AOL

**16-1** *The main menu for GEnie's Personal Finance and Investing Services.*

forums and the necessity of jumping around to different places to gather information is still a drawback there.

In sum, says Christy Heady, a syndicated financial columnist and coauthor of *Complete Idiot's Guide to Managing Your Money* (Alpha Books/Macmillan, 1995), "If you're an advanced financial whizbang, you really might love CompuServe. If you're a meek beginner, America Online might be your ticket." Christy started the relatively new Women and Investing section on Women's Wire, which includes a beginner's discussion section called Wall Street 101, primarily to help women learn enough to take charge of their own financial well-being. Women's Wire doesn't offer the research, analysis, and trading tools that the much larger services do, however.

# Yes, you can trade online. But should you?

Christy and others say that buying and selling through online brokers can save you money if you're already quite familiar with the markets and what you want, but do-it-yourself online trading isn't likely to cost less than deep-discount brokers. "Whether it's through America Online, CompuServe, or Prodigy, you can save more money in commissions from what you'd pay a full-service or discount broker," she says. "Online services are there to provide information, and the information is at your fingertips. Just make sure you don't make the mistake of not following up and monitoring those investments. It's like blindly writing out a check and never finding out what happened to your investment."

If you trade online, "You're not going to get your hand held," cautions Robyn. "You've got to know what you're doing in business, and what you're doing with your computer."

Says Becky Moores, a CPA and certified financial planner in Houston who uses both America Online and Prodigy's PCFN, Personal Computer Financial Network, "I still use an offline broker, but I'm the kind of person who doesn't need that advice. I can put in a trade on PCFN at midnight and it goes through the next day. There's something about hitting the Enter key yourself that's more satisfying than telling somebody over the phone to do it."

*Online services are there to provide
information, and the information is at your fingertips.*

*Tip:* Check out any brokers you're thinking of trading with, particularly if they aren't on a major service. The North American Securities Administrators Association (NASAA) at 202-737-0900 can give you the phone number of the agency that regulates securities exchanges in your state, and that agency can tell you whether a company is licensed to do business in your state.

PCFN, by the way, is an online discount brokerage service run by the Pershing Division of Donaldson, Lufkin & Jenrette Securities, and is also on eWorld. One of the other most popular services on Prodigy is Strategic Investor (see Fig. 16-2), which carries an additional charge.

```
 ┌─────────────────────────────────────────────────────────────┐
 │ ▟ Hot Stock Charts ▛   Investor's Business Daily   ** SAMPLE ** │
 │ CHAMPION ENTERPRISES (CHB)   AMEX  27.88  +   1.38            │
 │ SHARES: 7119      EPS RANK: 91      RS: 94     PE: 17         │
 │ WEEKLY                                              ┤┤┤├ ── 27 │
 │ HI/LOW                                         ┤┤├ ┤┤         │
 │ CLOSE                                    ┤┤┤├ ┤┤      ── 24   │
 │ 50-DAY                              ┤┤┤├                ── 21 │
 │ MOV AV                                            94╱   ── 18 │
 │ RS vs                                                  ── 15  │
 │ S&P 500                                                       │
 │    April      July      October     January     March        │
 │          (CHB) WEEKLY VOLUME in 000s                         │
 │  600                                                          │
 │  400                                                          │
 │  200                                                          │
 └─────────────────────────────────────────────────────────────┘
 [▤][◄][►][Menu][J][P][A-Z][X][Z][A][Pr][T][E]          [FREE]
```

**16-2** *A sample analysis from Prodigy's Strategic Investor service.*

# Research strategies: To each her own

"If I see something on Prodigy's Money Talk about a stock or find out about it on my own, I use Strategic Investor to look at its past history, its performance, and how it compares to other stock issues of like nature," says Judy Muldawer, who owns a computer products business in Albuquerque, and started the women's section on Prodigy's Money Talk bulletin board. "I also read the company news on Prodigy. If you track your stocks, on any given day there will be a highlight showing the date the company news was published. It might be about a new product or a dividend report or that the chairman has resigned or that they're merging, so it's often not real exciting, but it's all useful information. If you were to read the *Wall Street Journal* and comb it with a fine-tooth comb, you could get the same information, but it's so much easier to get it this way." (Also see the section at the end of chapter 12, *Custom news-search services.*)

Using Quote Track on Prodigy, Judy creates several portfolios, ones that give her daily updates on her actual investments and those that simply monitor stocks she's considering. "I put down that I bought one share at the price on that date, and then I look later to see whether it went up or down. That way I can watch the volume and how it's doing." She also downloads data from Prodigy directly into Quicken, her bookkeeping program on her own computer. Judy does her actual buying by phone, using her Schwab 500 account, however.

 *Tip:* On both Prodigy and CompuServe, and possibly some other services, you can download updates on your investments into Quicken or Managing Your Money personal finance management and bookkeeping software.

Christy Heady says she prefers America Online (see Fig. 16-3) because, "You can create a stock portfolio that shows how many shares you bought, and it will update it with the current trading price every 15 minutes. It's a real portfolio if you actually buy those stocks, but it's make-believe online." She, too, does her buying offline, through deep-discount brokers.

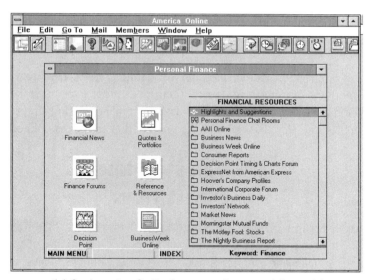

**16-3** *America Online's menu of personal financial management services.*

Becky Moores tried CompuServe, Prodigy, and AOL, and settled on the latter two. She uses both so she can double-check facts and trends, because she gets different information from each service, and they enable her to do different things. On Prodigy, for instance, she can create numerous portfolios to track both her actual investments and those for different groups of stocks or mutual funds she wants to follow and might buy if they look good after she watches them for a while. "AOL has room for only one portfolio," she says, "and you have to go to a separate business news forum, which I find very cumbersome. On Prodigy, I get the P/E (price/earnings) ratios and the highs and lows for the last 52 days, then, if I click on the details icon, it gives me the stock price and change from prior day. I use Prodigy mainly for P/E ratios, but just for the stocks that I own."

"For mutual fund information, Prodigy and AOL are about equal," says Becky. "On Prodigy you have to pay extra for some things, but my Prodigy bill comes out cheaper. I use it a lot, because the stock things are core. With Strategic Investor on Prodigy, I get all the data I normally have to go to the library for, such as earnings information per quarter or for the last year, or annual earnings for the last five years."

Like many of the most active members of CompuServe's Investors Forum, Robyn Green and her husband live off their investments, so they take it very seriously and have become experts. She uses CompuServe's Global Reports and Trends reports (see Figs. 16-4 and 16-5), among other resources online.

## How to use and not use investment forums

Having tried forums on various networks, Robyn swears, "CompuServe's Investor's Forum (see Fig. 16-6) is the best resource. The quality of the forum participants is head and shoulders above the other services. There are rules against solicitation and advertising, and that has had a great deal to do with the way the forum has developed. I think it sets a horrible tone for people who are trying to talk and learn, otherwise. It's hucksterism. When we see messages like that, we start asking those people questions."

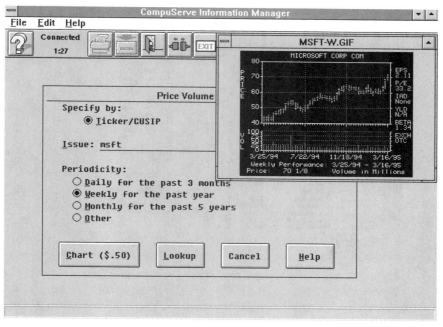

**16-4** *CompuServe's instant Trends analysis of Microsoft from March 1994–95, by week.*

**16-5** *Three separate reports from CompuServe's Trends analysis, for comparison with 16-4: trends over the last year, by week, for America Online, H&R Block (which owns CompuServe), and Microsoft. You can save each report as a graphics file on your computer for future reference.*

The offenders quickly get the idea that they're not welcome and leave for more fertile ground, she says. It's not just hucksters that get run out of the forum, either. Robyn says people in the Investors Forum give short shrift to people who just love to pontificate, make predictions, and tell others how they should

**16-6** *The types of files you'll find in the Demos and Information Library, one of the more than 20 libraries in the Investors Forum on CompuServe.*

spend their money, but who don't act on their own advice and risk their own money.

Robyn's advice about acting on tips from anyone in any forum, however, is candid and succinct. "Never. You don't know whom you're talking to," she says. "The value of the forums is getting ideas, which you then investigate on your own. Someone might tell you about a stock you never heard of before. You get that information, and go out and do your own research. It's a place to start. It's a source of information, as opposed to advice." She and others say you do learn to pay attention to comments from people whom you know are knowledgeable and trustworthy from their previous posts, however.

Christy Heady adds, "I go into these chat rooms and see beginners taking advice that's way above their heads. I would take everything with a grain of salt. If a doctor tells you that you have cancer, wouldn't you get a second opinion?"

Kathleen Capps, a Lieutenant Commander in the U.S. Navy, says, "The biggest benefit I've gotten from the CompuServe Investors Forum is a terrific sense of camaraderie and self-confidence to make my investments. . . . Many of the members are extremely knowledgeable, and the give and take among them is a fantastic education. In day-to-day life, there would be no way to meet and communicate with such a range of like-minded people. I interact with a wide variety of people online and offline, but rarely meet people who enjoy talking investments. Yet by logging onto the forum, I can instantly be in touch with hundreds of people with similar interests."

*The value of the forums is getting ideas, which you then investigate on your own. . . . It's a place to start. It's a source of information, as opposed to advice.*

"Actually, 'similar interests' is kind of a misnomer. The investment field is so varied that everybody has a different way of researching or making decisions. But it's exactly those differences that expose me to new ways of looking at things. Its a continuous growth process that takes me to new and unimagined places."

– 259 –

Kathleen is often at sea for months at a time, traveling worldwide or living abroad. So she uses CompuServe, because it has the most international connections of the commercial services, and telnets to Dow-Jones News Retrieval via the Internet so she can keep up with her investments by remote.

What she has learned in this process has given her a goal she says she probably wouldn't have thought possible without what she learned online. "Knowing all those people out there—many of whom are making a living doing this stuff—gave me the confidence that I could do it as well. I've always wanted the freedom to live and travel wherever I wanted in the world, but couldn't imagine a way that it would be possible. Now I *know* I can. All I need is access to a laptop, modem, and telephone, and I can follow my investments from anywhere in the world. I could live in Thailand or Athens or Istanbul and still conduct my business. Of course, I would first have to be at a point where I can support myself from investments, but knowing something is possible is the first step toward achieving a goal."

## What you get online vs. offline

"I'm a firm believer that online stuff is a supplement to other work you do," says Robyn Green. "Unless someone is strictly a technical analyst sitting there looking at charts, you need to listen to the news, read newspapers, and go to traditional sources of information. I read three newspapers a day and a handful of magazines every month. I can scan a great deal of information very quickly. When people use only what's online, their focus becomes very narrow."

# Paying bills and taxes, and spending what's left

## Online banking

Although Microsoft Network might do so eventually, the only service that offers even limited online banking so far is Prodigy. But their BillPay USA service is basically the same thing as CheckFree, an independent service that enables you to pay bills by modem (phone 800-882-5280). Either links to your bookkeeping software if you use one of the popular programs such as Quicken or Managing Your Money. The advantage of CheckFree over BillPay is that it can be set up for almost any bank, whereas Prodigy's BillPay is limited to about 20 different ones and costs more. Many local banks offer similar services, but they typically cost up to twice the $9.95 per month that CheckFree charges. You eliminate stamps, envelopes, and paper checks by banking online, and you save the time it takes to prepare bills the usual way.

If the security risks on the Internet are resolved—which merchants and bankers are working furiously to do—online banking might become more prevalent. Although most of us are now quite accustomed to banking by phone and ATMs, which use much of the same technology, it will be a long time before most people feel comfortable in letting their banking transactions filter through cyberspace.

# Tax time made easier

The IRS has been accepting tax returns filed electronically for several years, but never has it been easier than it is now. If you need a tax form at the last minute, you can download it straight from the IRS, on the Web, or from any of several other sites run by tax preparation businesses on the Web or on the commercial services (see Fig. 16-7). To accommodate people with different kinds of computers, they're available in a paper-emulation format known as PDF, which means you must have a program called Adobe Acrobat Reader to use them on your own PC. That program is usually readily available for downloading wherever you find the tax forms, and it's free and easy to use.

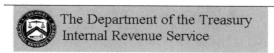

The Department of the Treasury
Internal Revenue Service

Peggy Richardson

Commissioner

- Tax Forms and Instructions
- Frequently Asked Questions
- Where To File
- Where To Get Help With Your Taxes

**16-7** *The folks at the IRS want to make it easier for you to pay your taxes, so they created a Web site. How thoughtful of them.*

CompuServe has an advantage over the other commercial services because it's owned by H&R Block, the tax-preparation company, which also owns one of the two most popular tax software packages, TaxCut. You can get either TaxCut or TurboTax online (for a fee, of course), and file your tax return directly from within that program, by modem. If you have a refund coming, that theoretically means you'll get the check within three weeks instead or eight weeks or more.

Tax forums are ubiquitous online, especially from January through April. Just use *tax* as a keyword search term, either on the Web or any of the commercial services.

# Online shopping

Each of the commercial services has an online shopping area, and the Web has several, with more likely to come. They range from little more than the equivalent of strip centers to what could become major mercantile zones where you can buy anything from furniture to coffee beans. Consider most of them still under construction, however, because they're a long way from what they'll be even a year from now. They're still too luxury- and computer-oriented in some cases, and have too much text either without photos or with blurry photos in others. Few of us, men or women, would buy something to reflect our personal taste when all we know about it is "Ralph Lauren towels, $7.99," or "red dress with calf-length skirt, $59."

As one woman put it, "I want to go to Spiegel's online and type in 'white blouse,' and see photos of every one they have." Right now, you can't do that, although you'll come closer to it on the Web than on any of the commercial nets. But that will all change as merchants get more sophisticated and confident about this new way of reaching buyers, and as the technology evolves and the fiber-optic cable transmission lines become widely available. Nonetheless, it's likely to take several years for online shopping to evolve into what some had envisioned.

CompuServe and Prodigy offer the most choices of the commercial nets, but America Online is expanding. All three offer or are soon to offer CD-ROM products that connect with their online shopping areas, and provide extras such as video and sound. The Internet Mall, Commerce Web, Downtown Anywhere, and the Internet Shopping Network (online counterpart of the Home Shopping Network) are among the many shopping centers on the Web. Not everything worth trying or buying online is in one of the mall areas either, because more people open up shop on the Web every day. So try various keywords, depending on what you're looking for, such as *mall, shopping, coffee beans, flowers*, whatever. Also see chapter 7, *Domestic domain*, about shopping for homes and real estate online; chapter 10, *Pleasurable pursuits*, about collectibles; and chapter 12, *100 ways to save time and money online.*

# Highlights: Online money management resources

## The commercial services

All of these services offer stock quotes that are updated every 15 to 30 minutes and access to specialized news sources or news wires, sometimes free and sometimes for an extra charge.

*America Online* The American Association of Individual Investors Forum, The Motley Fool, Morningstar Mutual Funds, Fidelity Investments, Investor's Business Daily, Business Week, Vanguard Group Mutual Fundation, Hoover's Company Profiles, and The Nightly Business Report. Also try *portfolio* as a keyword.

*CompuServe* The best way to start is to try both FIND finance and FIND investment, which will give you slightly different menus of what's available, and explore from there. The two main forums are Investors Forum and the forum sponsored by the National Association of Investment Clubs. Both will show up on a menu if you type FIND investment. Also, the Detroit Free Press Forum's Library 18, Personal Finance, has articles by financial writer Steve Advokat and other reporters, such as: "The advantages and disadvantages of electronic tax filing" (TX0216.TXT), "How inflation attacks your personal financial lifestyle" (IN0209.TXT), and "The financial struggle women face today" (FI0125.TXT). Use the command GO detforum. You can also find *Fortune* magazine and *Money* magazine's Mutual Fund Monitor; stock updates every 15 minutes; Standard & Poor's, Hoover's, and Dun & Bradstreet reports; a new service called Corporate Action and Municipal Notification; and much more on CompuServe. But many services come with surcharges ranging from $1.50 per report up to $60 an hour, billed by the minute. They can be cost-effective if you learn to use them judiciously, but pay close attention to pricing and instructions before you proceed.

*GEnie* Investors Roundtable, with sections for novices as well as a wide range for experienced investors, and the Daily Depositor report. Also GEnie Closing Quotes, Dow Jones News Retrieval, The Investment Analy$t, Charles Schwab Brokerage Services, SOS Investment Advisors, Dun & Bradstreet Company Profiles, Corporate Affiliates Research Center, Investment Reports, and TRW Credentials Service (also see Fig. 16-1). Many of these carry surcharges.

*Prodigy* Business and Finance Section —> Money Talk Bulletin Board. Within the Stocks section of the Money Talk BB, there's a section for women. "I'd say most of the women on the board are middle-aged, married, and financially secure," says Judy Muldawer, who started the women's section. "The issues that come up are things like socially responsible investments or taking care of elderly parents and how to keep them from being bamboozled. The typical question might be, 'My mother is widowed, she has $50,000 but no active income, and her broker has advised her to invest it in the following securities, or mutual funds, or treasury bills, or CDs, or whatever.'" Also check the HomeLife Bulletin Board, Quote Track, and Dow Jones News Retrieval.

*Women's Wire* Women and Investing section and Wall Street 101.

# The Internet and the World Wide Web

Good starting points to find other resources:

*The World Wide Web Virtual Library Finance Page* at the following address: http://info.cern.ch/hypertext/datasources/bysubject/overview.html

*FINWeb Home Page* at the following address: http://riskweb.bus.utexas.edu/finweb.html

*Finance Web Pages at OSU (Ohio State University)* at http://www.cob.ohio-state.edu/dept/fin/finance.html

*The Finance Page* at http//www.rain.org/~billman/finance.html

Other sites or services of interest are:

*The EDGAR Dissemination Project* Available on Pipeline and on the Web at http://www.town.hall.org/edgar/edgar.html; this service allows you to retrieve SEC documents or request that documents filed by specified companies be sent to you by e-mail. EDGAR is currently experimental, so still free at this writing, but it might not remain so. Increasingly, the Net and especially the Web part of it operate like drug dealers: give 'em a free taste until they're hooked, then threaten to cut off the supply if they don't cough up money immediately.

*Finance Net* Part of Vice President Al Gore's National Performance Review to cut government waste, at http://www.financenet.gov/

*Global Network Navigator Personal Finance Center* at http://nearnet.gnn.com/gnn/meta/finance/index.html

*Wall Street Direct* at the following address: http://www.cts.com:80/~wallst/

*Stock Market Data from MIT* at http://www.ai.mit.edu/stocks.html

*Hoover's Reference Press and Company Reports* at the following address: http://www.hoovers.com

*Security APL Quote Server* at http://www.secapl.com/cgi-bin/qs

*Nest Egg* at http://nestegg.iddis.com/

*Modern Portfolio Management Home Page* at http//www.magibox.net:80/~mpm/

*Fidelity Investments Home Page* at the following address: http://www.fid-inv.com/

*Baker & Co. deep-discount brokers* at http//www.cris.com/~bumm/baker1.html

Also see the Bibliography for references to *Net Money* and the AAII's book, *The Individual Investor's Guide to Computerized Investing.*

# CHAPTER 17

# Direct democracy & community involvement

When the students revolted in Tiananmen Square in Beijing in 1989, people all over the world cheered them on, and the students conveyed progress reports beyond China's borders, directly and daily, by fax. While the USSR collapsed in 1991, ordinary citizens of democracies worldwide rallied by e-mail, telling workers, scholars, reporters, mothers, friends, and strangers there, "We're with you, we'll help you, hang in!" And people there replied with daily updates, despite the Communist government's attempts to plug the flow. Since 1994, in besieged Sarajevo, e-mail via the ZaMir Transnational Net is often the only way people have to communicate with distant friends and relatives.

*Za mir*, by the way, means "for peace." If the medium is the message (as Marshall McLuchan is frequently misquoted), then the message of this medium is, "Let us make peace." Consider that in early 1995, while Mexico's economy teetered on the brink, the government was also contending with an uprising of peasants and farmers known as Zapatistas who banded together in and around San Cristobal, in Chiapas, to fight for their land. When the government launched a surprise invasion to quash their rebellion, Marcos, the Zapatista leader, immediately instigated a worldwide protest by fax. Perhaps it was coincidence, but the president halted his troops. Marcos reportedly carried a laptop computer in his backpack, even in the jungle, so he could get his compatriots' message out by e-mail. If you're skeptical, read this message from an anthropologist at an American university that was rapidly relayed through the Internet in mid-February by people from universities, churches, human rights organizations, and the media:

```
We've just received an emergency call from friends in Mexico. They tell
us that the Mexican army has surrounded the city of San Cristobal, in
Chiapas, and that the hospital in the nearby city of Comitan is flooded
with casualties. The press is being excluded from the area. The people
being attacked are the Mayan Indians and other poor farmers who've been
denied land and food since the conquest.

They've asked that we try to get word about this out via e-mail. While
we have no further information beyond this one call, I ask you to pass
this message on, or tell anyone you think relevant via any means so that
this does not occur in silence.
```

The alert turned out to be somewhat exaggerated, although not without foundation. But it resulted in reporters and human rights activists from other countries converging upon San Cristobal overnight. Mariclaire Acosta, president of the Mexican Commission for the Defense and Promotion of Human Rights, told the *L.A. Times*: "This is a real, postmodern war in the sense that information is the strongest ammunition of all. Use of the Internet has actually helped stop bloodshed. The distribution of information has made this one of the most low-casualty rebellions in history."

During any insurrection, dictatorships and authoritarian governments invariably move to cut off or control the media and other channels of communication. But the more people everywhere connect to the Net, the more impossible that will be. The downside to this is that we might wish we still had the media to filter out more of the misinformation for us, but things have changed irrevocably.

"The power of this medium is that citizens can now communicate laterally, without the intervention of the government or the media," says Barry Toll, a Vietnam veteran turned POW/MIA activist. "They can't keep so many things secret from us anymore. That is all circumvented now by individuals being able to say things to other interested individuals directly, bypassing all that."

*The power of this medium is that citizens can now communicate laterally, without the intervention of the government or the media.*

– 269 –

Barry learned about that power firsthand. During the Vietnam war, he was in the U.S. military, assigned to a special team under the Secretary of Defense. His job made him privy to highly confidential information, through which he deduced that some kind of secret deal had been made between President Nixon and the Vietnamese regarding war reparations, and that it had resulted in POWs and MIAs being, in his words, "abandoned by the U.S. government."

More than 15 years later, two things happened that changed his life, and many others' too. First, the Soviet Union crumbled in August of '91. The following month, the Senate coincidentally created their Select Committee on POW and MIA Affairs. By then, it looked like The Cold War was all but over, so Barry itched to tell what he knew, now that what he wanted to divulge wasn't likely to jeopardize national security in the way it could have before. But first he wanted to find out what others knew too, and whether they thought it was smart to talk or if it could be harmful to the country or relatives of those who were never accounted for, and listed as dead or MIA.

"So I thought CompuServe would be a way to find my former compatriots," he says. "First, I went to the main membership directory. I sat there all night one night, just pumping in names from the past. I finally hit on one, so I sent him a message. Then I typed in FIND Vietnam, which led me to the Military Forum. I searched the forum's membership directory, and found out he was a member there.

"Next, I used a few search terms from my unit in combat and my special forces unit in Vietnam—acronyms like MACV, SOG, and LRRP that no one would be likely to use unless they were involved too. Then I typed Ranger, POW, and MIA, and punched SEARCH. I got hits! Indeed, the hits were people whose reputations I knew. I immediately sent one of them a message."

Within days, Barry heard from several others who had been in Vietnam the same time he had, some of whom had even served in the same unit he had. Over the next several months, they constantly compared notes, and put more pieces of the puzzle into place. Meanwhile, the Senate Select Committee hearings were getting nowhere, which distressed the families especially, and fueled a heated debate in the Military Forum.

"Alex Humphrey, one of the sysops, finally told me, 'You must go forward. This will be the last, best chance to get the truth.'

"So I was the first witness who came forward and revealed that Henry Kissinger, who was National Security Adviser at the time, and President Nixon were informed by intelligence people that hundreds of POWs were being left behind at the end of the Vietnam war. . . .

"One of the first things the Committee asked me was when I started speaking of these things publicly, and to whom. So I told them: 'On CompuServe, to members of the Military Forum.' They were astounded."

Since Al Gore became vice president, which catapulted his long-time advocacy of the national information infrastructure front and center, and Newt Gingrich took up the banner when he became Speaker of the House, the Senators would probably not be so amazed if that hearing were held today. The President, Vice

President, and most members of Congress now have e-mail addresses. Congressional bills under consideration are routinely uploaded to Thomas, the Congressional site on the Web (http://thomas.loc.gov), and the White House distributes information online daily.

Clinton and Gore set the pace when they took their election campaign online through electronic town meetings, and many have followed suit since when running for local and statewide offices. During her campaign for the state legislature in the fall of '94, Colleen Burkett set up a branch of her campaign headquarters on Cyberia, a BBS based in her hometown of York, Pennsylvania (see Fig. 17-1). In the last month of the campaign, she too held weekly town meetings. Online. It was her first time as a candidate, and she was a Democrat running against the Republican incumbent in the year of a history-making Republican victory at that, so she lost. But taking her campaign online garnered Colleen an Associated Press wire story, and established her as a forward-thinking leader, which is exactly the image she wanted to convey. Whether she ever runs for public office again or not, she'll have more clout in her state and community hereafter.

**17-1**  *Colleen Burkett's campaign forum on Cyberia for her campaign for state office in Pennsylvania.*

# Think globally; act locally

Even though your fax or e-mail message can help affect outcomes nationally and internationally because of the cumulative effect, as demonstrated by some of these examples, if you're more interested in what you can do locally there's nothing more effective than freenets. These are free, local BBS systems modeled on The Cleveland Free-Net, the Seattle Public Network, Santa Monica's PEN, and similar networks throughout the U.S., Canada, and beyond. They're free to users, although not to operators, so they're usually funded by a combination of local companies, governments, and universities. Many have set up computer terminals and training in public libraries and community centers to provide access even to those who don't have computers. In fact, the PEN network was originally established as a way for people to communicate about how to help the homeless, and homeless people had direct input into the plan they developed.

Through a local freenet, you can update your car registration, sign up for a Parks and Recreation class, or just get to know your neighbors. You can also mobilize people at election time, tell city officials what you think about agenda items posted for public review before the council meeting next week, and conduct all kinds of meetings without meeting. Some are connected to the Internet too. If you'd like to find out how they work and how to start one in your community, join the Communet mailing list on the Internet by sending e-mail to listserv@uvmvm.uvm.edu. Leave the subject line blank, and in the body of the message type: SUBSCRIBE communet *yourname*.

As more and more people get online, there's also a growing trend toward creating regional- and local-interest gathering places. Women's Wire, for instance, has sections for all geographic regions in the U.S. and some for Canada. The Internet has numerous places by city, region, state, and country. They're usually identified in some part of the address by either an airport code—such as PDX for Portland, Oregon—a commonly used abbreviation for a city, such as SF for San Francisco, or by postal codes, such as CA for California. If there's already one where you live, try a few logical abbreviations and you'll probably find it.

# Nonprofit organizations on the Net

Many nonprofit organizations, particularly those involved in the arts and human services or social action, now use e-mail. However, few are taking advantage of the many online opportunities beyond that. "The barriers to NPO access to the Internet include the cost of hardware, software, and network connections; lack of computer or network literacy; lack of appropriate technical support; and high turnover rate for staff and volunteers," says the FAQ (frequently asked questions) for the soc.org.nonprofits newsgroup. "There is also the reality that most people are not only not on the Net, but they also see this new technology as merely a fad—and such people usually make up the majority on NPO boards. Board members hold the purse strings, and they can be particularly hard to convince when it comes to the need for basic office supplies, let alone Net access." So if you're a volunteer or board member, help change that. And teach, don't just do things for them, so they won't be dependent upon you and can accomplish things when you're not available.

In fundraising and marketing for the arts, health agencies, and social service organizations over a span of 17 years, I learned that many nonprofits, particularly those in the social services, have a poverty mindset that keeps them poor. As the quote from the FAQ indicates, the board members and volunteers are just as likely to think this way as staff. Most of them would never run their own businesses or even their households with the handicaps they expect nonprofits to endure daily. Therefore, way too many are still saying "But we can't afford to be online" when they could be on the Internet for an average of $17.50 per month, according to *PC Magazine*'s 1995 report on the nets. Nearly all online services, including Pipeline and Netcom, offer free software, Netscape is less than $50, and Internet in a Box under a hundred, street price. That's a one-time investment. All things considered, surely that makes this the most cost-effective medium available on the planet, used resourcefully. So many are still operating under misconceptions. What they really mean is "We don't know how."

– 273 –

"How" is you take the next 250 bucks that comes in and buy a used computer if your office doesn't already have one. Even an old one will do to get started, but it should at least have a hard drive. Better yet, get it donated. There are national networks now that funnel donated computer equipment from individuals and corporations to wherever it's needed. Take whatever kind of printer you can get, at first, but when you invest in one get at least an inkjet, because it will cost you only about $50 more than a dot-matrix and the results look as good as a laser printer for correspondence. If you spend $100 more, you can get a laser printer and save a lot on outside artwork and layout charges. If you can get a computer new enough to run Windows or a late-model Mac, so much the better.

Next, sign up with the local Internet-access provider that comes the most highly recommended in your community. They might even donate an account. If not, with auto-navigation software that lets you do most things offline, you're still talking less than $200 a year. Even from remote areas where there's no local-access number, you can get a SprintNet hook-up for less than ten bucks a month with some services. So much for the "We can't afford it" argument. Would you use a table knife for a tool after someone invented a screwdriver? Would you still type letters on a manual typewriter or run flyers off on a mimeograph machine? The fact that you're even reading this book says probably not. Besides, you'll make and save that $200 many times over by what you'll learn and the connections you'll make online. If you need further convincing or assistance, the following guides come highly recommended: *Guide to On-line Systems for Nonprofits* and *The Care and Feeding of On-line Networks for Nonprofits*, both available from CompuMentor, phone 415-512-7784 and e-mail cmentor@well.com.

Despite my harping on how social activists could do so much more with this medium, there are enough creative and resourceful examples to fill a whole book, so let's look at a few of the ways nonprofit organizations are using online services and ways they could be:

## Online newsletters

ChoiceNet, among others, uploads a weekly newsletter to relevant Internet mailing lists and newsgroups, which enables everyone who gets it to distribute it widely. Volunteers upload it to libraries on commercial services, for instance, and incorporate the information into their local publications.

## Get-wired programs

With funding from the Joyce Foundation, the League of Women Voters created their Wired for Democracy initiative. They now run their own section in the Political Debate Forum on CompuServe, where staff members and volunteers lead discussions and answer voters' questions. The League also uploads their voter guides far and wide through the same methods. The California, Columbus, Ohio, and New Jersey Leagues are already online, and they hope to get all local Leagues online in their respective communities eventually. "This is really just the beginning," says League spokesperson Matt Farrey. "Gradually, the grass-roots organizations are getting online. We just have to get them trained on how to use it."

If your organization is national, you might be able to get a section in a forum on a commercial network or a Web site. For the forums, ask the chief sysop. For Web sites, find a compatible location, and either ask the site owner and operator or contact The Internet Nonprofit Center for recommendations.

## Affordable access assistance

The Telecommunications Cooperative Network in Washington, D.C., also a nonprofit, has set up private forums for nearly 50 organizations—usually national ones—on various commercial services, such as CompuServe. Many of these forums are private, thus out of public view, but members nationwide can log on and communicate with each other, hold real-time conferences, upload files to the library and to one another, and send e-mail to anyone on any other network. TCN also negotiates discounts for regular accounts because, as a membership organization, they wield collective buying power. They'll do the comparison shopping and negotiating for you, and recommend the best service for your needs based on the nature of your organization, what you want to do, and who's willing to offer you the best deal. You must join TCN to get these benefits, but they can get better rates than any single organization could.

## Regional theaters

There's a growing number of regional theaters that have carved out a corner for themselves on America Online, according to Ann-Marie Miller, the development director at McCarter Theater in Princeton, New Jersey (which won the 1994 Tony Award for Outstanding Regional Theater, she'll have you know). The area would be hard to find normally, as is the case with many of the best things online, because it's in the *New York Times*' @times forum. People from Center Stage in Baltimore, Hartford Stage, and others show up regularly to discuss ticket policies, ticketing software, fundraising, and other common interests. Ann-Marie says being online has also given her direct access to a state legislator to make McCarter Theater's voice heard among the outcries over the assault on arts funding. "There have been several messages posted on a number of mailing lists and newsgroups about national advocacy efforts," she says, "so I've been able to obtain e-mail addresses for all legislators currently online to share with board members, staff, and audience members."

Through the soc.org.nonprofits newsgroup, which she was able to join even though it's on the Internet and she's on America Online, Ann-Marie also found the information she'd been seeking on how to apply for one of the U.S. Department of Commerce grants to assist nonprofits with information technology. The reason she wants to apply for that grant is noteworthy in itself: "We have a [local area] network that connects nine regional theaters in New Jersey, and we're sharing audience demographic information to get a more accurate picture of people who attend cultural events. We're looking for funding that will enable us to obtain software and data analysis expertise."

## Community revitalization

The Rockefeller Foundation funded the Millennium Communications Group to produce a report called "Communications as Engagement: a Communications Strategy for Revitalization," which they've now uploaded to their own Web site. It includes a browsable listing of more than 235 model programs. The Web address is http://cdinet.com/millennium.

## Fundraising and board development

Heather Newman, a reporter at the *Tucson Citizen*, produced a major report for the paper called "Who Runs Tucson," using tools and techniques that journalists call computer-assisted reporting, better known as CAR. Anyone who needs to raise money, recruit board members, or find out who's swaying public policy behind the scenes could do the same. "All you need to know is how to use a basic database program," says Heather. Actually, you need to know quite a bit more, as she concedes, mainly about how to find and download information from the Internet and government agencies, then how to get a database or spreadsheet program to translate the voluminous data you retrieve into something meaningful and useful. But it's do-able.

"There are very few records that are open to reporters that aren't open to average citizens," Heather says. "The only difference is that your average citizen hears 'no' and thinks that's the end of it. Or they don't understand that the information is available and you have a right to see it; whereas reporters make a big noise about it if they have trouble getting access to public information. A lot of government offices have been dragged into the electronic age kicking and screaming, so this information doesn't come to you all wrapped up in a nice package. It's often just raw data, and the only recourse is to learn to use programs designed to crunch all this data down into some form you can understand. It's not easy.

"I generally use the Internet to find out who's got it, then pick up the phone to find out how to get it. Even if the data isn't available online, you can usually get it on disk. When I was working on this project, I had the county tax assessor's office sort out the top 100 pieces of residential property in the county. Those are the people who have the money to give or have the clout to call other people."

For the curious among you, the essence of CAR is pulling comma-delimited data from someone else's database into your database or spreadsheet program. The software programs reporters use are the same ones sold at any software store, such as Paradox and Excel. Sometimes you have to get the data on disks in person, but much of it is now available by modem and more will be steadily. Even local government records are increasingly accessible because they're going online too, and you can telnet directly into their computers and find what you need. But that's the boring part. It's after you collect the data and start to analyze it that it gets interesting. It helps to understand why if you get more of the details from Heather:

"I used Paradox to track every time a person's name was mentioned during interviews, surveys of local community folks, readers and staff, and from lists of boards of directors and power committees and groups around town," says Heather. "I ended up with close to 5,000 entries—more than 900 different names. I used Paradox to organize these into ranked lists: who got mentioned the most, who got mentioned the most for a particular category (top five, up-and-comer, and so on), who got the most 'points' if I weighted the categories they were named for and such. We used these lists to help select our top folks. I used computers to analyze the female vs. male membership ratios of boards and committees encompassing more than 1,000 people. I also used computers to figure out how often women were winning political offices here if they ran. And finally,

I used the computer to compile statistics from three of Tucson's biggest employers to see what types of jobs women and minorities really get in this town."

Just think what you could do with that kind of information for fundraising, educational outreach, and political campaigns, and for coming up with a hot board prospect list. There are probably things you could do with it that the media haven't even thought of yet, because they don't think from your perspective. Most reporters are still new to CAR research and analysis; they learn how to do it from each other and through two Internet mailing lists in particular: IRE-L, for a nonprofit organization called Investigative Reporters and Editors, and CAR-L, which is the computer-assisted reporting list.

Other helpful resources for this kind of data collection and interpretation, says Heather, are the Investigative Reporting section in the Journalism Forum on CompuServe; the FOI-L Internet mailing list, which is about how to get government information under the Freedom of Information Act; and the NICAR-L mailing list, from the National Institute for Computer-Assisted Reporting. These are all discussion groups, not just lists. Some are called lists merely because the discussions are distributed by automated group e-mail.

"One of the great things about the CAR sections is that a lot of people are so proud that they can do what they can do that they're willing to help anybody who needs help. There are really good, content-rich, usable suggestions . . .. And absolutely the best source for usable data is the National Institute for Computer-Assisted Reporting. They take it off nine-track tapes that a lot of us don't have the equipment to use, and convert it. They often go to the trouble of dividing the information up by state and zip code, and clean up data that has problems, then sell it at cost."

## Specialized networks

Many nonprofits say they value the specialized networks, even though they're not as big as the Internet nor as slick as the major commercial services. The most popular one for the social service organizations is HandsNet. And for environmental groups there's EcoNet, which is part of IGC (the Institute for Global Communications), the same organization that runs PeaceNet and ConflictNet. Their address on the Web is: http://www.igc.apc.org/index.html.

For the arts, ArtsWire is a popular one (on the Web at http://www. tmn.com./oh/Artswire/www/awfront.html). Like HandsNet, it's a separate service. Some independent networks, such as HandsNet, actually cost more than the mainstream services, but it's the concentration of knowledge and expertise that people appreciate.

"I've been quite critical of HandsNet because of their cost and isolation from the Net," says Terry Grunwald of the North Carolina Client and Community Development Center. "But I've come to believe that it offers tangible benefits that cannot be found anywhere else. First, HandsNet uses professionals in the field to provide core program and policy information that nonprofits need. The information is abstracted, edited, and annotated, which reduces the need to sift through the reams of garbage on the Net. Most NPOs simply don't have the time to surf and play around. The filtering function is of great value. Besides, most of the information simply cannot be found elsewhere, especially in areas such as substance abuse, and for children, youth, and families who have paid information providers. HandsNet responds to the needs of their users—and not with what techies think that NPOs need, which is ideological debate and chat."

# World Wide Web

First a word about why the Web is such a "big deal" in the opinion of net aficionados. America Online, CompuServe, and Prodigy offer full Web access, either through their regular access numbers or with Web browsers built into their software, and Delphi, eWorld, and GEnie intend to catch up by the end of '95. These gateways are a good way to explore at first, but if you find the Web as irresistible as most of us do, the smart move is to sign up with an Internet service provider that gives you direct access so you can avoid the higher online time charges of the commercial nets (see appendix B). Many services, such as Pipeline and Netcom, provide free Web browser software that also enables you to use gopher, telnet, FTP, and e-mail. The catch is that the software won't run at less than 9,600 bps, nor would you want to try the graphics on the Web at a slower rate, and eight megs of RAM and a hard drive are minimum for PCs. If you've got that, then the Web is the easiest and most fun way to use the Internet, so it helps get people who are technophobes over the hurdle and online.

Although it's still in the early stages of development and will grow and change a lot over the next two to ten years, the Web is already a rich resource and has enormous potential as a way to reach people, conduct research more easily, and connect with others in your field in order to learn from and help one another. If it's a matter of outreach rather than management interests, you can also do that well on the commercial services, as the League of Women Voters has begun to do. There are already far too many useful resources to list here for either purpose. However, as one who believes that the arts are as necessary to nourish the soul as food is to nourish the body, it's disappointing that these organizations aren't included in the key sites (although there is an arts management mailing list). The focus on the Web is almost entirely on human services, with some on the environment and other issues.

## The Foundation Center

This site is mostly background information, although it does include their weekly newsletter on news of foundations. No, you can't search the Foundation Directory or Foundation Index databases here, alas. They're not feeling that charitable, apparently. The Web address is http://fdncenter.org/.

## Guide to Internet Resources for Non-Profit Public Service Organizations

This is part of the excellent Subject-Oriented Clearinghouse on Internet information at the University of Michigan. You'll find hundreds of text-only documents and references to other resources. Go to http//asa.ugl.lib.umich.edu/chdocs/nonprofits/nonprofits.

## The Internet Nonprofit Center

This is a very well-done, comprehensive site (see Fig. 17-2) that can easily serve as one-stop shopping for donors and potential volunteers, as well as nonprofits looking for other organizations and help on the Net. The Heliport includes links to virtually all of the nonprofit organizations on the Internet, and the Gallery contains annual reports and brochures for many of them. There's also information on more than 8,000 nonprofits, plus helpful documents for donors, such as the Best Buys for Big Hearts list of organizations recommended by the American Institute of Philanthropy, which sponsors The Internet Nonprofit

*The Internet NonProfit Center*

Home to donors and volunteers

Gallery  Library  Parlor  Heliport

Gallery
of Organizations

Library

Parlor

Heliport

**17-2** *The Internet NonProfit Center, the most comprehensive collection of information tand links to other resources, http://human.com/inc on the Web.*

Center. There's an automated system for nonprofits to post notices of volunteer opportunities, and for people interested in volunteering to respond to those.

*Tip*: Even if your organization isn't on the Web, you can send volunteers-wanted notices by e-mail to Cliff Landesman, who operates the Center, and he'll post them for you. His e-mail address is clandesm@panix.com. To get to the Center on the Web, type http://human.com/inc/ in your Web browser.

The following are also highly recommended, and can be reached through their links to the Internet Nonprofit Center or their own addresses: America's Charities, which highlights member organizations and offers assistance and screening information to potential donors; ReliefNet, which is a way to find out about the world's current crisis zones and how to help through any of several international relief organizations, including the American Red Cross, Oxfam, CARE, the American Friends Service Committee, and the International Rescue Committee; EnviroWeb from EnviroLink (shown in Fig. 17-3), the largest concentration of environmental information and links on the Web; Philip A. Walker's Table of Contents, which is a directory to other information, including online United Ways, plus business and financial information for managers and board members of nonprofits.

Welcome to the EnviroWeb, a project of The EnviroLink Network. The EnviroLink Community is one that constantly grows through the work of all of its volunteers, including its users. Find out how you can add new documents to the EnviroWeb and help with already existing projects. If you have any comments or questions, please send mail to the EnviroLink Network's administrators.

**17-3** *EnviroWeb from EnviroLink, the largest concentration of information and links to environmental resources online, http://envirolink.org on the Web.*

## The Meta-Index for Non-Profit Organizations

Also known informally as "Ellen Spertus` Web site" because that's who created and maintains it, this has the most comprehensive list of online nonprofits, with information about each of them, including the Women's Environment and Development Organization. The 45-page directory covering mailing lists, newsgroups, gophers, and Web sites, called *Internet Resources for Not-for-Profits in Housing and Human Services*, is also available here. Use your Web browser and go to: http://www.ai.mit.edu/people/ellens/non/online.html.

# Elsewhere on the Internet

*Newsgroups* The two newsgroups that draw the most people are soc.org.nonprofit and alt.activism. The former is usually helpful, but the latter tends to frequently degenerate into haranguing.

*Mailing lists* The fundraising-oriented ones are the most popular among nonprofits, big surprise. These are: Gifts-L, which is about planned giving; Prospect-L, where people talk about prospect research; and Fund-L, which is mostly university development folks hobnobbing about how to get major donations, according to Cliff Landesman of The Internet NonProfit Center.

For information on how to subscribe to those lists, ask in one of the related newsgroups or check the Lists of Lists or the List of Publicly Accessible Mailing Lists (see appendix G).

# On the commercial services

The best resources for nonprofit organizations on the commercial nets are, by far, Women's Wire, where many women-related nonprofits have their online homes in the Organizations section, and The WELL, which is also a socially conscious place with solid information and a lot of well-informed people. If feminism is your cause, another good place to be is on Echo, in the *Ms.* magazine conference.

As for the major commercial nets, GEnie has the Public Forum/Nonprofit Connection Roundtable, but it's overrun by soapboxers. CompuServe's Issues Forum has a Nonprofits section too, although traffic and knowledge are both light. On Delphi, check the list of custom forums for The Path to Peace. In general, however, you'll do just as well cruising the forums on the arts, politics, or other relevant topics to connect with people interested in what you're working on. There are worthwhile nooks, such as the one in @times on America Online, mentioned previously, where regional theater people hang out. But, frankly, the big commercial services aren't where it's happening if you're looking for help as a staffer, board member, or volunteer for a nonprofit organization. And because most of the NPOs that are online are on the Net, the commercial services are useful mainly just as gateways.

# Political information and debate forums

Here, too, the information could fill much more than a chapter. One of the most intriguing places is the Political Participation Project (see Fig. 17-4), which, by involving people who visit this site on the Web, intends to answer the question "Can interactive media improve political participation?"

**17-4** *The Political Participation Project on the Web, at http://ai.mit.edu/project/ppp/home/html.*

If you're content to just debate politics rather than get involved, there's ample opportunity everywhere online. The Internet tends to be a free-for-all, which might appeal to you, but try the soc.politics newsgroup first because it's moderated and people there can give you a good reading on which of the alt.politics groups are worth considering, if any. Most newcomers, particularly, will feel less overwhelmed if they try the commercial services first. Not that people don't get flamed there, but at least the sysops act as referees if things get out of hand. Each of the services has one or more forums or sections on politics, so just use *politics* and *government* as keywords to find them. The Republican National Committee and Democratic National Committee each have their own forums on CompuServe, and you can find lists of e-mail and earth mail addresses for federal elected officials on the Internet (http://policy.net is one choice), as well as on most of the commercial services.

The majority of people who use online services at present are white, well-educated men with good incomes. If only the upper strata of society can afford access to online networks and only neighborhoods with the right demographics get the fiber-optic cable and computers in schools, libraries, and community centers, we'll have widened the gap between the haves and have-nots even more, with all attendant consequences. Right now, the have-nots include women—whether by choice or chance—in all strata, not just the lower classes.

This is an issue, therefore, that should be a concern of anyone interested in equal employment and educational opportunities. If you'd like more information on what the realities are or how to get involved, join the Computer Professionals for Social Responsibility newsgroups (comp.org.cpsr.talk or comp.org.cpsr.announce), and read what the U.S. Government's Information Infrastructure Task Force has to say about universal access and gender and electronic networking in the documents at their gopher address, gopher.iitf.doc.gov.

By the time this book is updated, I hope community, social, and political activists have learned to use all the variations of online resources and opportunities as resourcefully as they've learned to use photocopy machines, typewriters, and telephones. Online communication is really nothing more than just another tool. But it's likely to turn out to be the most powerful and effective one since the invention of television and the telephone. As the saying goes, use it or lose it.

# CHAPTER 18

# Online research basics

**W**hen Esther Gwinnell, a psychiatrist in Portland, Oregon, heard that Medicare was going to require doctors to file all their payment claims by computer, she knew she had to enter the era of the modem, like it or not. But she didn't realize that filing forms electronically would turn out to be not only a great convenience, but truly mundane compared to what else she could do with that modem.

Like other working women, Esther was already writing in the margins of her calendar just to cram in everything she had to do each week. One day not long after installing her modem, while talking over this common dilemma with a colleague at Oregon Health Sciences University, she wondered aloud how she was going to find time to see all her of patients, handle her teaching load, process all the paperwork, and manage things at home, yet still find time to do the tremendous amount of research she needed to do for her upcoming testimony in a court case. "Paperchase, on CompuServe," was her friend's instant answer. He explained that Paperchase is IQuest's name for Medline, and Medline is a way to electronically search the 8.6 million articles in the archives of the National Library of Medicine in Washington, D.C., in a matter of minutes. Esther couldn't wait to see if it was as good as it sounded. It was.

"I find it wonderful to get information from the library without leaving the office," she says. "It's almost miraculous. I type in a few keywords, and ten minutes later I've tapped the world literature on the subject! I got quite giggly and excited the first time I used Paperchase."

Later, Esther learned that she could get a $30 software program called Grateful Med, which was created specifically for direct access to Medline (from the National Technical Information Service, phone 800-423-9255). Grateful Med automates much of the search process, and enabled Esther to access Medline directly at about one-third less than she was paying through the Paperchase gateway on CompuServe. But since Paperchase reduced their rates to $18 an hour, it costs the same to access Medline directly. Because Paperchase is menu-driven, meaning you can choose your search strategy from onscreen menu items and you can choose only one search criterion at a time, it's a lot slower process than using Medline directly if you're a skilled online researcher. But Esther prefers the menu-driven searches because they're easier, so she still uses Paperchase.

*I find it wonderful to get information from the library
without leaving the office. It's almost miraculous. I type in
a few keywords, and ten minutes later I've tapped the
world literature on the subject!*

# What you can find online

You can find all of the following online, and more: articles, abstracts, and citations from magazines; newspapers; highly specialized newsletters and journals;
research reports; financial data and analyses; corporate and government data
and reports; information from all kinds of research volumes and directories,
such as *Books in Print, Who's Who*, and the *Directory of Associations;* and
theses and dissertations (although those are mostly on the Internet). In sum,
just about anything that's in print, and some that isn't.

The goal of this chapter is not to make you an instant expert, but rather to give
you an overview of what online research is like and a few tips so you'll feel
more confident about trying it yourself. Here are two of my favorite examples:

## When it's good, it's very good

For a magazine article on what was real vs. what was fiction or fad in the latest
nutritional claims, I once needed to find the most authoritative information I
could, as quickly as I could, about developments in nutrition during the previous year. I lived in Santa Cruz, California at the time, which has a pretty good
library for such a small town, and a university library too. I spent an entire
afternoon and at least ten dollars in quarters for the photocopy machine at the
library, yet came home with little more than a list of publications that would
probably have what I wanted, but that weren't available locally.

The best chance of finding such specialized information, they said, would be to
drive an hour-and-a-half to San Francisco, to a med school library. I knew that
would end up taking all day and most of the evening, because I was incapable of
going to San Francisco and doing nothing but making a quick trip to the library. I
had never tried online research before, yet found exactly what I needed in

ZiffNet's Health Database Plus on CompuServe, in addition to a government report that I hadn't discovered in the library citations. I downloaded two articles, at $1.50 each. Altogether, it took 11 minutes and cost $6.40, including the charge for the two articles. Rates have decreased since, so at current rates that same search would cost only $5.13.

## And when it's bad

Another time, knowing that one of the keys to online research is making your search terms as specific and narrow as possible, I thought I had a guaranteed success because all I needed to know was more about Juneteenth. One word—how much more specific can you get? I knew that Juneteenth was the day the slaves in Texas found out they were free, a year after the Civil War had ended, more or less.

I also knew I'd find something if I searched the Texas daily newspapers, so I searched the *Houston Post*. That part went fine. Then I went into some section of one of the more expensive databases—without reading any instructions, mind you—and got lost in a very lengthy document that mentioned Juneteenth precisely once, in reference to names of holidays. Because I hadn't read the instructions and hadn't used this database before, I didn't know how to halt the search. That's how I got into real trouble. In nothing flat, I ran up a bill of $99. Resourceful creature that I am, I talked the service into a partial refund and sold all the useful information I'd gathered to a Juneteenth Celebration planning committee. So it was a lesson learned, and I was fortunate that I came out of it so well. To this day, I can't keep myself from scanning instructions when I enter a database, even if I already know them by heart.

# A get-oriented guide to online research

All online databases intended for research are set up and operate in much the same way. Think of it as walking into a huge library, where the databases are like special rooms or wings with different names. Because this is an electronic library, everything is small enough to store in file cabinets, each of which also has a different name. Inside the file cabinets are the files, which is exactly what the separate online documents are called too.

Although what you'll see onscreen is just plain text, without pictures to click on as you might be used to in other areas online, if you know how to read a menu in a restaurant and pick something from the entrée section and the dessert section, you can do online research. Apply the same principle and you'll be impressing your boss, your friends, and yourself from day one.

If you have only an occasional need to research something online, you'll do well to choose the menu-driven search option rather than the advanced-user option, because you get explanations and the opportunity to request instructions with the menu method. However, if you find yourself using online databases frequently, you'll save time, ergo money, if you take a class or do more reading to learn the actual syntax and commands that are hidden behind the menus. Here's what the screen for those choices looks like in Knowledge Index, the after-hours discount version of Dialog available only on CompuServe:

```
Select one of the following options:

       1 Menu-Assisted Searching
       2 Command/Advanced Searching
       3 How to Use KNOWLEDGE INDEX
       4 Database Descriptions
       5 General Information

Copyright 1995 Knight-Ridder Information, Inc. All rights reserved.

Enter option NUMBER and press ENTER to continue.
 /H = Help                     /L = Logoff
```

# Four key concepts common to most searches

It's helpful to know the four key concepts that apply to almost any online database search: defining your search terms, Boolean logic, widening the search with wildcards, and narrowing the search parameters.

## Defining your search terms

Start by writing down all the words that come to mind for what you want to find and what forms you'd like to consider, such as reviews, abstracts, or articles. It could be something like "recent uses of technology in manufacturing cardboard boxes" or, with the example used elsewhere in this book, "cause of and treatment for fibromyalgia." Then cut out all the propositions and conjunctions, which would make it "cause treatment fibromyalgia." You could start with just *fibromyalgia*, as I did, and if that gets too many "hits" (results, or a list of items found), narrow it by the other two terms in each succeeding step—assuming you're doing a menu-driven search rather than a command-line search, which is more complicated. Medical journals might use the word *etiology* rather than *cause*, so you have to consider all alternatives. Write down what you're going to use first, second, and third, or as far as you want to go in homing in on exactly what you want to know.

## The basics of Boolean

Most databases use an algorithm known as Boolean logic, also sometimes referred to as AND, OR, NOT search syntax. Fortunately, most of the database services make it pretty simple, particularly if you use the menus.

In its simplest form, you're looking for words that fit your search criteria or could fit, but saying "Ignore things that might look right, but aren't." To put it another way, AND, OR, and NOT specify what to include and exclude.

➤ AND means get only the articles that include both this word *and* the other word.

➤ OR means get anything that includes either this word *or* the other word.

➤ NOT means get the article if it includes the first term, but *not* if it includes the other term.

For example, if you're researching retirement options and you type just IRA as a search term, you'll get everything about the Irish Republican Army as well as individual retirement accounts. But if you type IRA NOT Irish Republican Army, you'll get information only on the money kind of IRA. There are also ways to link, further define, and exclude more words by grouping them together, but once you understand the basic concept, you'll understand how that works too. Because there can be minor variations depending on which database you're using, such as whether it requires square brackets, [], or parentheses, (), let's leave that to the pros to explain. It's usually described in the instructions when you log into an online database.

## Widening your search with wildcards

Wildcards are symbols that mean "and any other string of characters." The wildcard character is usually either an asterisk, *, or a slash, /. For example, if you're searching for something about computers but it could be found under either computers or computing, you would use comput* because that would find either term. This can make your search too broad and lead to too many hits, but it's good to start broad so you won't miss possibilities you didn't think of, and you can keep narrowing the search until the number of hits you get is manageable.

## Narrowing the focus of your search

Depending on what the database allows, you can narrow your search by keywords or subject, date or a range of dates, publication name, author name, and whether the words appear somewhere in the text or must appear in the title or headline to qualify. The trick is to keep narrowing your criteria until you get no more than, say, 100 citations, read through those offline, go back online, and request more detail on just a few. Some databases give only citations on the first pass, some give summaries or abstracts, and some allow you to download the full article for as little as $1.50; others will only fax the article to you, usually for $8 to $16. You can use these articles for reference, but not reprint or reuse them or distribute them in any form, because they're copyrighted.

# The same search on four different services

You want to choose the service and the database within it that will give you the most useful information for the least money, of course. However, that's not always obvious just by reading about the rates. As you saw in chapter 8 when you looked up *fibromyalgia* in the free HealthNet database and got nothing, free can be just another word for useless. Watch what happens when you use *fibromyalgia* as a search term with four separate services. To keep things as simple as possible, all but one of the following services are available on CompuServe, which has the best research resources of the commercial nets.

# Health Database Plus

(On CompuServe, GO HLTDB or GO HealthDB) As you saw in chapter 8, Health Database Plus includes consumer magazines, newsletters, journals, and publications from health organizations. Only abstracts are available for most of the journals, but you can get the full text of most magazine articles. It also incorporates relevant full-text articles from publications that aren't specifically about health. Health Database Plus is updated weekly, but includes only data for charts, not the graphics.

This is the next logical database to try, after HealthNet, because you pay only regular CompuServe charges of $4.80 an hour or eight cents a minute. That's considerably less than the $24 an hour or 40 cents a minute for Knowledge Index, which you'll try next. The catch with Health Database Plus is that you pay $1.50 each for any full articles you decide to download.

There's a free support forum, however. To get to it, choose Database Customer Service from the Health Database Plus menu. There, in Library 10/HealthDB, you'll find a user's guide (HDPUSR.TXT) as well as a file that lists all publications included in the actual database (HDPPUB.TXT). You can also find those files by using CompuServe's File Finder feature (GO IBMFileFinder or GO MacFileFinder). It pays to download those and read them offline to find out if the publication you want is there, if it's a particular one you need, and how to plan your search on paper, offline, so you can save money while the toll counter starts ticking during your actual search.

*Tip*: Health Database Plus works the same as Magazine Database Plus, Computer Database Plus, and Business Database Plus.

Because the term *fibromyalgia* defines exactly what you want to find and is only one word, it's not likely to lead to false hits (unless you disregard the instructions, of course). False hits are what you get when you use IRA as a search term and end up with hundreds of hits related to both the Irish Republican Army and individual retirement accounts. Fibromyalgia is a good search term because it adheres to the first rule of successful online research: always define your search criteria as clearly and concisely as possible. Therefore, there's less risk of pitfall

#1 in online research: inadvertently running up a big tab by not planning and defining your search carefully beforehand.

In Health Database Plus, you can choose to search by keywords, subject headings, publication names, publication dates, or words in the article text. Start with any one of those, then narrow your search by any of the others if you turn up too many hits on the first try. Fibromyalgia as a keyword results in 74 hits, or citations. Because that represents everything on the topic from among well over 200,000 articles, it's best to narrow the search to get only recent references, not only because new studies could mean new knowledge, but to save time and money. Narrowing the search to get only articles published after 01 Jan 94 instantly cuts those 74 citations to just eight. So far, so good.

That's so few, however, that it's worth choosing Display Article Selection, which lets you see citations of all eight articles, but here's where pitfall #2 of online research comes in: It's so easy and automatic that you end up downloading a story or three, even though you promised yourself that you'd only window-shop. Unless you exercise real willpower, you'll rack up charges quickly. During this sample search, "Fibromyalgia: a misdiagnosed & misunderstood syndrome" from *Executive Health's Good Health Report*, "Antidepressants for fibromyalgia: Are they effective?" from *The Back Letter*, and "Comparison of patients with chronic fatigue syndrome, fibromyalgia, and multiple chemical sensitivities" from *Archives of Internal Medicine* all proved to be irresistible—or seemed so at the time, at least.

This entire search and the three article downloads took only four minutes. With three downloads at $1.50 per, it cost $4.50, plus the connect time of 32 cents. That's a lot of legwork saved for less than five bucks. Not only were the references and articles themselves useful, but they gave names of researchers as well as research and treatment centers, and stated that the Arthritis Foundation is likely to be able to provide much more information, most of it probably for free. So Health Database Plus is clearly a good investment, all in all. It could be an even better one if you didn't read everything while you're online, as I did, but simply captured the whole search during the process and read the entire file offline. Then you can decide which things to download once you've seen the whole picture and your perspective is less influenced by impulse. (I did that too. The file for all three of the searches described here was 142.5K, which is a lot to sort through.)

# Knowledge Index

(GO KI on CompuServe) Just to make sure what you've seen so far is a good picture, if not the whole picture, it seems prudent to go ahead with the Knowledge Index search anyway. It costs $24 an hour, or 40 cents a minute, including all other charges for time online. As you saw previously, a well-defined and well-planned search can take only a few minutes. Even though Knowledge Index costs more than either of the other two, this is a discount-priced subset of about 100 of the more than 400 separate Dialog databases, so it still costs substantially less than a search on Dialog or Nexis would. The following is a glimpse of what the main screen looks like:

```
                KNOWLEDGE INDEX(R)  Main Menu

              Knowledge Index Database Sections

Choose a subject CATEGORY by entering an option NUMBER.

    1  Agriculture                14  Food (one file)
    2  Arts                       15  Government
    3  Biology, Biosciences & Biotech.  16  History
    4  Books  (one file)          17  Legal Information
    5  Business Information       18  Literature & Language
    6  Chemistry                  19  Magazine
    7  Computers & Electronics    20  Mathematics
    8  Corporate News             21  Medicine
    9  Drug Information           22  News
   10  Economics                  23  Psychology
   11  Education                  24  Reference
   12  Engineering                25  Religion
   13  Environment (one file)     26  Social Sciences

Copyright 1993 DIALOG Information Services, Inc.  All rights reserved.

  /L = Logoff      /M- = Previous Menu      /MM =  Main Menu
```

If you follow a similar process of reading search instructions first, then choosing the Medicine section and telling Knowledge Index to search all relevant databases, you get to Medline. In just seconds, Medline tells you that there are 752 articles on fibromyalgia. Gulp!

No problem. Simply choose to narrow the search, then limit it to the current and last publication year, which immediately culls just 64 articles from those 752. Reading through all 64 citations online and requesting abstracts for about 20 of those articles took only 13 minutes and cost $3.55, total, plus the connect time of about a dollar. Better yet, this inquiry cost even less than Health Database Plus, despite the latter's lower rates, and the abstracts give descriptions and results of medical studies, so Knowledge Index resulted in the most information for the least money. Remember at the beginning of this chapter when I said that the service that costs less might not always be the best bargain or the most cost-effective? Case in point.

 All three searches took less than half an hour, combined, which is less time than it would take you to get to the library, much less do the same research there. And you couldn't do it there anyway unless your nearest library had all these publications or allowed free use of Medline, both of which are highly unlikely.

*Tip*: Some libraries offer free use of online databases. InfoTrak is the most commonly available and it can be helpful, but it's not as sophisticated as the databases covered here. The few libraries that have Nexis usually won't allow you to print things out because it costs considerably more, so it's best for quick research where you can write down results. Some libraries even offer free searches by librarians who know how to find just what you're looking for fast. If the one near you does, count your blessings.

## Medline on Nexis

The Knowledge Index search in the previous section was done in Medline, but there are several ways to get to Medline. So I did a similar search in Nexis to compare. At a basic rate of $39 per hour or 65 cents per minute, Nexis is one of the most expensive, as well as one of the best, online databases. The charge for each database on Nexis varies, however, and their rate for Medline is $24 an hour, the same as Knowledge Index. Using *fibromyalgia* as the search term again turned up 57 abstracts, and cost $25.65 for the time online, and downloading all 57 of those abstracts to disk at 2.5 cents per line was an additional $20.94, whereas on Knowledge Index you can simply do a screen capture and pay nothing extra to save the abstracts you find. The cost of the Nexis Medline search and saving the abstracts to a computer file came to $46.59 altogether.

The Medline information was so technical in both cases that it was difficult for a layperson to understand. A search of Nexis' Current News database, which retrieves references from only the past two years, proved to be a better strategy. A Nexis service adviser suggested limiting the search to articles where the term *fibromyalgia* was mentioned at least three times in each story, so the stories would more likely be about fibromyalgia, rather than merely containing a passing reference to the word. That search resulted in 101 hits at total cost of $33.07, yet the articles were from good sources, including *American Family Physician*, *British Medical Journal*, *Archives of Internal Medicine*, and the *Atlanta Journal and Constitution*.

A better strategy, still, might have been to search just the *New York Times* for articles from the last two years, which would have cost only about $10 on Nexis. You could also search just the News or Magazine Index databases in Knowledge Index, but that would require two separate searches and, although they're great for some purposes, they're limited to the popular press, so you would have missed some of the articles that were the most useful on Nexis, yet not have saved much money.

## IQuest's Paperchase version of Medline

It turns out that Esther was right. At $18 an hour, Paperchase (GO PCH on CompuServe) charges the least for Medline, yet provides menu-assisted searching. I had to get help from one of the pros the first time, though. Because she knew exactly how to do the search, it took her only about two minutes to come up with just over 100 abstracts, which would have cost only 60 cents. Add a little for fumbling-around time and false starts, assume you capture it all on your hard drive while you're doing the search so you pay nothing extra for downloads, and you'd still spend less than five bucks.

*Tip*: It's important to stress that online research, as wonderful as it is, can get very expensive very fast unless you really watch what you're doing, particularly on something like Nexis or Dialog, or even IQuest. Unless you're on an expense account or will be reimbursed, stick to the services available through the commercial networks. Nexis is by subscription only, but will allow you to subscribe to just a small subset or one database.

# Saving money on online research

As you've seen, online research can get expensive, especially if you don't bother to read the instructions or don't do preliminary research to decide on the most cost-effective information source. After you've done a few searches, however, you'll know which service to use for which purpose. Doing research online also saves you countless hours, and can make you as excited about your new skill as Esther was when she first used Paperchase. The more you know about how to search and where and the more tricks you learn, the less it costs. The key to saving time without spending heart-stopping amounts of money is to plan your search very carefully before you get into the section where the meter's running, grab the citations you find to read offline while you're not connected, and go back online a final time to get only what you want. To save you flipping back to chapter 8, here's a rerun of how that process breaks down:

1 Download all the descriptions and instructions first, read through them offline, and decide which database sounds like the right one to use.

2 Read through the instructions about how to define and refine your search terms, then write them down on paper, keeping them as specific and as narrow as possible.

3 Go online, set your capture buffer to save the whole search to a file in your computer (often the PageDown key toggles it on and off, depending on your software, or look on the menu bar while you're online for Capture), do the search, tag the citations as quickly as you can, and log off.

4 Read through the citations offline and choose just the most important ones. You'll be much more judicious about this when you can see the entire list of what you found in step 3 in front of you, offline, than you will if you ignore this process and tag what you want while you're online.

5 Have your list highlighted and handy, go back online to grab the articles you want, log off again, and read through them leisurely, offline.

A few final tips:

➤ See the previous chapter, 17, for more ideas on ways to do research electronically, and places online to learn how journalists do what they call computer-assisted reporting (CAR).

➤ CompuServe has many other research databases (GO reference, for starters) and GEnie also has several excellent ones as well (see their About Research and Reference Services), including Dow Jones News Retrieval, access to Dialog, and the Trademark and Tradenames databases.

➤ If you'd like to learn more, one good source is *CompuServe Companion: Finding Newspapers and Magazines Online* from Bibliodata, which is listed in the Bibliography. You can order it through the CompuServe store (GO order).

➤ If you like the idea of online research but would rather somebody else do it for you, contact the Association of Independent Information Professionals (also known as information brokers or online researchers) at:

245 Fifth Avenue, Suite 2103
New York, NY 10016
Phone 212-779-1855, fax 212-481-3071
73263.12@compuserve.com

Their online home is in section and library 4, Information Professionals, in the Working from Home Forum on CompuServe (GO work).

# CHAPTER 19

# "Senior doesn't mean old"

It seemed like any ordinary day when Ilene Weinberg logged on to SeniorNet, on America Online. But Ilene has Parkinson's disease, so she seldom knows when she'll have difficulty doing today what she was able to do with ease yesterday. In the middle of reading messages from her many SeniorNet friends, the words on the screen suddenly blurred and her whole body felt like it was shutting down.

"I was hallucinating and couldn't type straight," Ilene recalls. "I thought I'd had a stroke. I typed 'Please listen to me. I'm in trouble.' All they knew was my screen name, but one person knew I was always talking about the weather in Boston. I don't know how they found out who and where I was so fast, but the Boston police called me and asked, 'Are you the lady with the computer who's in trouble?'"

The medication Ilene was taking had caused the problem, not a stroke, thankfully. It was partly because her spell, as she now calls it, hit while she was at the keyboard that Ilene called for help online rather than by phone. But it was also because her SeniorNet friends are just as much a part of her life as any local friends or relatives are, and she knew they would help her when she needed them.

*I'm 68 and in a wheelchair a lot of the time because of the*
*Parkinson's. But when I'm online I'm 37, tall, blonde,*
*athletic, and ready to roll!*

"It's absolutely amazing," she says. "The support we give each other is nothing short of fantastic. When somebody loses a mate, they often come online to talk it over with people there first, before they even tell their kids. When somebody's despondent, we cheer them up. When it's your birthday, you have a hundred people telling you 'happy birthday.' Good news and bad is transmitted to the whole bunch. I've got a new family. It's delightful!"

Before joining SeniorNet, Ilene didn't even know how to type and says she was so technophobic that she was reluctant to get a Cuisinart, much less a computer. It was her son who talked her into trying SeniorNet. Being online has brought her not only a new group of friends but a new attitude, with many more opportunities and activities than she would normally be able to participate in, all right at her fingertips. "I'm doing a lot of writing in a relatively structured way with a group of writers," she says. "We have to write every week and we critique each other, so it has improved my abilities and given me a support group to do the writing I've always wanted to do.

"But my cyber-romances are the most exciting part. I'm 68 and in a wheelchair a lot of the time because of the Parkinson's. But when I'm online I'm 37, tall, blonde, athletic, and ready to roll!" She jokes about it, but here's how Ilene expressed her real feelings to one of her male friends by e-mail:

"The connections we make through this strange new medium seem to have a special quality. How else can one explain the almost mystical closeness we feel for one another through a keyboard and our fingers? It seems to create a bond almost greater than that which we feel for those near and dear to us. It is a communion of souls connected by a need resurrected from our earliest childhood memories; the psychologists call it unconditional love.

"In the past several weeks, due to crises of my own, I have had an opportunity to experience this phenomenon. This is what I have felt for you and this is what I have received from others. Sandwiched in between the joshing on the one hand and the florid expressions of love on the other lies a genuine kernel of caring that gets lost in the growing-up process, and that is the greatest nutrient for the growth of a soul. It is a special, quasi-religious bond we have with each other.

"Although my words are often ponderous, they contain much truth. We are granted this special thing, and we should react to it gratefully. Let it support you as it has supported me at times that we might falter. With this love may we have the courage and fortitude to return to each other all the good stuff that makes this turn on earth worthwhile."

*The support we give each other is nothing short of fantastic.*

Lest you think that Ilene's experience online is unusual, here are a few excerpts from comments from others. They're from an online discussion thread started by members about how SeniorNet has changed their lives:

Subj: Re:SeniorNet has changed my life
From: Iam Zephyr

I've never been a social butterfly, but SeniorNet has changed that for
me. Now I feel as if I want nothing but to be able to fly from one place
to another to meet all the wonderful people I've "met" online. I've
never been the recipient of so much generosity and thoughtfulness, and I
am hard-pressed to learn how to accept it. Where would I be without
SeniorNet? I would be unseen, unheard, unknown, unloved, as I was
before. Now I am inspired again—writing poetry, collecting poetry from
others, exchanging letters with interesting people. I hadn't used a
dictionary since high school, now I have to keep one handy to be sure
I'm not using the wrong words to express myself. I used to sit glued to
the TV watching boobs bob; now I am glued to my computer watching Bob's
booboo (just kidding ;-). I have cried and laughed. I have agonized and
been exhilarated. I have felt lost and lonely, and come away found and
full of hope. I haven't done this much thinking since the first time I
"turned my brains off" so I wouldn't die of boredom. Never before have I
known so many people who share the same feelings, the same ideas, the
same ideals. Never before have I felt as if I were truly a participant
in the human condition rather than just an observer. If I could only
physically reach out and give back the warmth and love I have felt
coming to me from the people here on SeniorNet. — Zephyr


From: Texas M549

I shall change the word "changed" [my life] to: enriched, educated,
amazed, inspired, and most of all, accepted.

You have added laughter, fun, loving, caring, a sense of belonging, a
true sense of friendship, and the list goes on and on. I still have the
feeling, even after one year, that when I go "into" the chat room, we
are truly all there together. I hope I never lose that feeling. Just
think about sitting in your own house and talking to people from one
side of the country to the other; it truly boggles the mind. I have
watched people struggle with pain, loss, and grief, and still come
through it as beautiful human beings who are truly an inspiration to me.
A fine example of what can be, if we so choose. I have felt such a sense
of friendship and reaching out across the miles.

I have met people who in this "real world" I would never have known. I
have learned to accept them as my friends, no matter how our beliefs and
lifestyles differ. And I wouldn't change a thing about them. The
knowledge and intelligence so many of you have and the subjects you
discuss and how well informed you are re: politics, world conditions,
etc. is enviable. I am anxious to meet more of you "face-to-face."

Love to each one of you. — Jane


From: Glorious

I like your comment about SeniorNet having ADDED to your life. Being of
an age in which we live with losing friends, family, and sometimes our
independence, it is wonderful to be able to find something that adds to
the enjoyment of life. My life has been great—good family, great kids,

good friends. But the new ones I have met here have been a blessing, and one unexpected at this time in my life. It's like being able to amble down the street and chat with your neighbors on a summer evening, or over the back fence after you have finished hanging the wash or raking leaves....

How times have changed! Our grandchildren grew up with Sesame Street, and we are growing old with SeniorNet. — Glo

From: SaJanina

This cyberspace connection has enriched my life deeply. It has brought new, warm, loving people into my life whom I might not otherwise have had the chance to meet because of our geographies. Some friendships have developed offline as well, which is even more wonderful. And then there are the forums. For better or worse, I find that the ideas presented give me the opportunity to work out again in my mind my own positions on issues and to formulate ways to represent those positions. That is, after all, excellent mental exercise!

Most important, though, is that I feel very privileged to have had the opportunity to get in on this phenomena at this time in my life. I'm hopeful that five years down the road I will still be enjoying the warmth and kindness of many of the friends out there in the air. The creation of a community in cyberspace is a very exciting concept, and I am delighted to be a part of it. It is an extra blanket on a cold night. It is a cool breeze on a warm one. — SaJanina

# You're never too old to learn new technology

Being about to turn 50 myself and none too enamored of terms like *senior*, I stopped into SeniorNet's chat room one morning to ask for ideas from members there about a title for this chapter. Most said they were glad to have reached the stage of life where they'd earned the right to be considered senior. "Senior doesn't mean old," replied Pat Decker. That sounded like a fine title in itself, and so it is.

When I asked if there was any hope of getting my active 77-year-old mother online, they said sure, and promptly mentioned several SeniorNet members who are in their 80s, and even beyond. When I was 40 I'd never used a computer and was sure I'd never learn, so I can understand how someone could think she was too old to learn how to use a computer or communicate online. But take it from the SeniorNet members and me, it's just not true.

Even fingers with arthritis or rheumatism or hands afflicted by carpal tunnel aches or twinges can click a mouse and tap keys most days, because the range of motion required is minimal. To use the easiest-to-learn services and SeniorNet on America Online, specifically, it's better to have a PC that can run Windows or a Mac, with a mouse or trackball for either kind of computer. But AOL and CompuServe and some other services still have DOS versions of their graphical-interface software, so you don't necessarily need the extra memory that Windows requires. You can get started on a used computer with a 40-megabyte hard drive, probably for less than $350, or a new computer for around $900.

Desktop models are easier on older hands than laptop or notebook models, because the keyboards on the laptops can be rather cramped (unless money isn't a concern and you can pay $3,000 or more for IBM's notebook model with the full-sized, collapsible keyboard). The advantage of the laptop is that you can take it with you if you're traveling, and use it even in a car or an RV. You'll also need a modem, of course, but AOL says many of their members still use 2400-bps modems, and friends or relatives would probably be glad to give you their old ones when upgrading to a faster model. Or you can get a higher-speed one for about $50. You don't need a second phone line either, although it's convenient so you won't miss calls while online. But most people get by with just one line. (See appendix A.)

SeniorNet is an independent, nonprofit organization that trains people over the age of 55 to use computers, and hosts a forum on America Online (see Fig. 19-1). There are SeniorNet learning centers in cities throughout the United States and in New Zealand (see the listing at the end of this chapter). If there's a SeniorNet learning center or computer training program in your city or nearby, they'll help you get started, as will your computer-savvy children, neighbors, and friends. So will the tech support people of the online services themselves—and for free—at least as far as how to install their software and find your way around online.

Increasingly, community centers and libraries have computers available for public use, too, and you might lobby them to make arrangements with SeniorNet to provide access for those who don't have or can't afford comput-

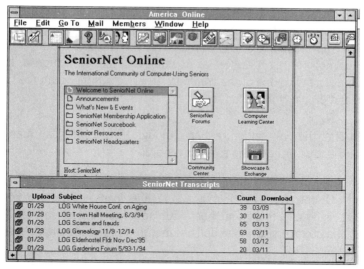

**19-1** *SeniorNet, on America Online, opened to a listing of archived discussion files in one of the libraries to show some of the topics members talk about.*

ers. SeniorNet is an excellent way to get started, though, if you can get your own computer. Members get unlimited access to just SeniorNet on America Online, plus an hour on the rest of the service. Many members also maintain separate accounts on AOL so they can also access other forums, yet stay within the AOL allotment of five hours a month for less than $10 on the rest of the service.

*Tip*: AARP is also on America Online, and AARP members save a buck a month on AOL's basic monthly fee.

One of SeniorNet's interesting sections is their oral history program, called Generation to Generation. It links with AOL's Kids Only Forum to provide an easy way for kids to ask older adults about what things were like during World War II or the Depression or just to get their perspectives on various aspects of life. Students use this connection for research for school papers, to learn about the past in a way that's more interesting to them than history books, and to gain insights into events and eras through details that history books rarely include.

Whether you go the SeniorNet route on America Online or join another commercial service, there are many, many people online who love to help newcomers, and free sections where you can learn what's where and how to do everything from posting messages to downloading files to using the Internet. It's easier to start with one of the commercial services rather than the Internet, however, and probably more fun too.

SeniorNet isn't the only option, either, although it's the only one that offers a discount for access. Among the alternatives are:

➤ AARP (American Association of Retired Persons) also operates sections on America Online and CompuServe, although its focus is more on social and political issues affecting its members. Anyone over 50 is eligible to join AARP, and members save one dollar a month on America Online's monthly fee. (See Fig. 19-2.)

➤ CompuServe sponsors the Retirement Living Forum, and has a section called Village Elders within the Issues Forum.

➤ Delphi's Custom Forum 211 is called Singles Around 50.

➤ Prodigy has the Seniors Bulletin Board.

➤ Women's Wire offers a forum called The Older Women's Network.

There's no reason to relegate yourself to only the "senior" forums or sections. That's just a good starting point, and a gateway to a whole new world online. Remember what Pat Decker says: "Senior doesn't mean old." Read it and believe it!

# SeniorNet Learning Centers

Annual membership is $35 for individuals and $40 for couples. For $9.95 a month, SeniorNet members get unlimited access to SeniorNet on America Online, plus one hour for the rest of the forums. For more information, write to:

SeniorNet
399 Arguello Boulevard
San Francisco, CA 94118

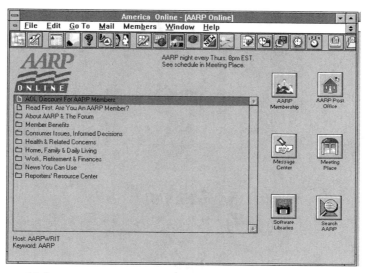

**19-2** *AARP members get a discount on America Online's monthly fee. They're also on CompuServe.*

or call 800-747-6840 or 415-750-5030. Here are the locations of all the SeniorNet Learning Centers:

# Alabama

Huntsville: 205-880-7080

# Arkansas

Hot Springs: 501-622-1802          Little Rock: 501-660-4110

# California

Bakersfield: 805-327-8511          Sacramento, northeast: 916-485-9572
Culver City: 310-202-5855          San Francisco: 415-771-7950
Fullerton: 714-526-2775            San Francisco: 415-922-7249
Huntington Beach: 714-960-7671     San Jose: 408-448-6400
Oakland: 510-531-9721              San Mateo: 415-377-4735
Orinda: 510-254-5939               Santa Cruz: 408-429-3506
Sacramento: 912-264-5462

# Colorado

Colorado Springs: 719-685-5721

# Florida

Fort Myers: 813-334-5949
Ocala: 904-629-8351
Sunrise: 305-742-2299

Tampa: 813-974-5263
Winter Park: 407-647-6366

# Georgia

Atlanta/Smyrna: 404-801-5320

Savannah: 912-651-7559 or 351-4520

# Hawaii

Honolulu and Oahu: 808-845-9296
Honolulu, Kokua Outpost: 808-528-4839

Kahului and Maui: 808-242-1216

# Illinois

Peoria: 309-682-2472

Springfield: 217-525-5699

# Indiana

Indianapolis: 317-849-1099

# Kansas

Overland Park: 913-469-8500, x 3844

# Kentucky

Lexington: 606-255-2527

# Louisiana

Baton Rouge: 504-923-8025

# Missouri

St. Louis Gateway: 314-530-2933

# Nebraska

Omaha: 402-552-2359

# Nevada

Las Vegas: 702-363-2626

# New Jersey

Ewing: 609-883-1009

# New York

Valhalla: 914-785-6793
New York City: 212-636-6782
Forest Hills (Queens): 718-699-1010
Hudson Guild (Chelsea): 212-924-6710
Stanley Isaacs Neighborhood Center
 (Upper East Side): 212-360-7620

Kingsbridge Heights
 (Bronx): 718-884-0700
Staten Island: 718-981-1500
University Settlement
 (Lower East Side): 212-473-8217

# North Carolina

Wilmington: 910-452-6411

# Ohio

Akron: 216-867-2150

# Oklahoma

Oklahoma City: 405-728-1230

# Oregon

Eugene: 503-345-9441

# Pennsylvania

Philadelphia: 215-276-6148

# Tennessee

Nashville: 615-329-8963

# Texas

Dallas: 214-768-4332
Houston: 713-963-4151

Nacogdoches: 409-564-2411
Waco: 817-666-6154

# Vermont

Essex: 802-878-9530

# Virginia

Springfield: 703-922-2474

# Washington

Bellevue (Seattle area): 206-232-5892

# Washington, D.C.

202-362-9292

# Wisconsin

Appleton: 414-735-4864

# New Zealand

Wellington: 022-64-4-382-3127

# CHAPTER 20

# Women belong online

This is probably going to sound like heresy, especially in this book, but a lot of women and some service providers are ambivalent about places online especially or exclusively for women, mainly for five reasons:

# The rationale against online areas for women

> Women's services online (and off, for that matter) can become places of exclusion rather than inclusion, or "pink ghettos." Among people in the online services industry, "community" has become a buzzword when talking about what women want online, but it's a term that has been co-opted and devalued. Yes, people—both male and female—do indeed enjoy and want a sense of community online, which is precisely why the commercial services are going to continue to hold their own, contrary to predictions that the Web will eventually obliterate them all. But what men in charge of planning online services too often mean lately by "community" in relationship to women is that they think what we really want is a backyard fence environment where we can chat and gossip to our hearts' content. Balderdash. And flat wrong.

> Depending on what tone is set from the beginning and how well or whether they're staffed, women's sections sometimes devolve into nothing but trivial chitchat or, worse, whining, grousing, and anti-male polemics. It happens because many of the women who would be more likely to steer the conversation and content in another direction avoid women-only sections. They're not online just to chat, and they identify more with their professional peers or other people with whom they have something more in common than their gender.

> Most women don't like being lumped in a catch-all category. The usual assumption about a section for women online is that it's either a bastion of strident feminists or a place to discuss only home-and-heart concerns. Very few of us want to be stereotyped or stuck on either end of that continuum.

> Even though it's not true that women's sections invariably have a feminist focus, the subject does come up. When it does, if it's a public section it's inevitable that a certain type of man emerges out of nowhere with comments typical of this one from a discussion on America Online: "I'm

beginning to think that all the ladies on this board are suffering from permenent [sic] PMS. Why don't you girls find yourselves a nice guy, take a few cooking classes, and settle down & get married? Then you can all have lots and lots of babies and find out what real fulfullment [sic] is all about. I know this is just a phase you girls are going through but come on, snap out of it. By the way how come feminists are always so angry? (I of course exclude lesbians from this question—they should be angry at themselves.) P.S. I love all women, even the apparently screwed up ones on this board."

➤ Finally, there are typically accusations of sexism when women-only forums and sections are created, and there's some legitimacy to that complaint. There are those who advocate in favor of these types of forums on the grounds that the online environment is hostile to women, therefore women need to be protected. More balderdash. Anytime someone in authority tells me they're going to create something for my protection, I'm immediately suspicious. Most of us fend just fine for ourselves, thank you.

# Why there should be places especially for women anyway

The flip side, of course, is that there simply is a different dynamic when men are present and when they're not, and women certainly are likely to discuss some things—from periods, to relationships, to office politics—when they know they're out of the range of male ears. Or eyes, as is the case online. And I don't see *Men's Journal* or *Woman's Day* folding in the interest of being politically correct. So I've come to the conclusion that if places online for women persuade more women to get online, then it's good that they exist. We're all free to either participate or stay away.

More important, I can't forget that in the toughest crises of my life, it has always been women who reached out and helped pull me through, every time—usually just by listening and being there when I needed them to be rather than trying to fix the problem for me, as men tend to try to do. I must also confess that I'm glad such enclaves exist, and there are times when I rely

on some online women's section rather than cope alone or discuss my dilemmas more publicly.

For instance, when I was still new to New York City and found a lump in my
breast, it was the women in WIT on Echo who helped me find an excellent
doctor and get an appointment the same day. When it was apparent that I was
likely to lose the job that brought me to New York all the way from California,
and was really worried about being in that most-competitive place, knowing no
one, they were the ones who helped me accept the inevitable, find the confidence to stay, and make it on my own. Echo feels smaller and more private
than something like CompuServe, because it simply is much smaller. I tend to
post concerns of such a personal nature as those in a private section for women
only, like WIT, whether on Echo or a far bigger service.

Closed forums or sections for women aren't the only answer, though. The public ones, where men can also participate, create a nonthreatening, almost
anonymous opportunity for men and women to try to understand each other
better and to discuss some of the role changes that are affecting all of us today.
Minds begin to open and attitudes slowly begin to change, which leads to messages like this one in the Women's Forum on CompuServe from Meredith
Lynne Huestis:

"I must add, I am so pleased to find males on this forum, and males
comfortable in being males and in discussing their issues vs. females'.
What a tragedy if this were entirely a male-bashing forum. All of us,
particularly those of us 40+, are going through a transition more
difficult than any young female or male can imagine. I was brought up—as
were most of my counterparts—to believe in the male over female
supremacy. . . . One raised in the atmosphere that I was raised in can
feel only tenderness toward males at this gender crossroad, and
sympathize with the havoc it can play on them emotionally too."

Another factor favoring women's sections and services is that they're increasingly being started by women for women, finally. This, too, has its pros and
cons, predictably. As the online services and advertisers have become increasingly aware that they're missing out on selling things to the person in the
household who makes most of the routine buying decisions, there's a big push
to get more women online. Along with that comes a lot of people positioning
themselves as experts in the same way everyone last year was an instant

Internet marketing expert, and a lot of start-up services that might not last. But nobody starts any kind of business without having some sincere passion for it because it's too rough a ride, so I can only wish them well.

# Pioneers in online services for women

There are local BBSs run by women throughout the country, and many women have been online since the mid- to late-70s. But there are a few women who have been activists in getting other women online. They have set the pace and the standard by starting their own services or, more recently, creating sections on major commercial networks.

## Echo, The East Coast Hang-Out

Stacy Horn, who launched New York City's Echo in March of 1990, was one of the first women to build an online service of her own from the ground up. She has a master's degree in interactive telecommunications, and is also one of the cofounders of WON, the politically oriented Women's Online Network, which influenced the direction she took from the start. Because she consciously created it as a place hospitable to women, Echo has always had more female members than most services.

"I actively recruited women," says Stacy. "Whenever they got online I would go out of my way. If men don't like a place, they complain. If women don't like a place, they just leave. To counteract that, we send each woman a letter asking if she would like to be assigned a mentor. The mentors help them not just with the commands, but the culture too. We contact every single user." All the mentors are volunteers and women. However, only about 25 percent of the new women on Echo take them up on the offer, even though Echo requires manually entered commands rather than the easier point-and-click commands. That's one more indication that it's not primarily technophobia that keeps women from going online, as some have claimed.

Although she might not be conscious that she's doing anything different from what other sysops do, another thing Stacy does to cultivate the atmosphere she wants is that she's constantly online, out front, and often outspoken. Although these comments are from a phone conversation between us, they're typical of her personality, which is consistent with her online persona:

On *Wired* magazine: "It's so boy-oriented. *Wired* is about gizmos and golly-gee-whiz, look at all this cool tech stuff. It's so shallow. It's made women think [being online] is geekier than it is."

On the Electronic Frontier Foundation mentioned in the Preface: "I hate the whole frontier imagery. This is a frontier of the mind, maybe. We're intelligent, educated, sophisticated, cultured people. True frontier people are not online; the kind of people online are the kind who get manicures."

On being online: "If you're not wired or in some way reachable by e-mail, you're going to fall out of so many informational loops that are available online," Stacy advises. "You used to have to go to luncheons and cocktail parties to network. Now you have to be online. . . . When you talk about networking information loops and job loops, it's all about power. If you're not using online networks, it's like saying I'm not going to use the telephone."

The effect of Stacy's high profile on Echo, through responding personally to members' e-mail and participating regularly in discussions, is that Echo had a distinctive character from the beginning. It isn't infused with her personality, nor would it fall apart if she sold Echo and left the fold, because she has laid a foundation strong enough to maintain the framework on its own. Part of the way she did that was to make sure that half the sysops are women. Another way has been to keep the price low enough that Echoids, as they call themselves, log on frequently. Many think of it as part of their daily routine and feel a strong sense of belonging to the community. It's so much of a community that it can feel rather cliquish. Not that they're cold to newcomers, but unless you log on and participate frequently enough for people to get a sense of who you are, it's easy to feel like an outsider. Echo has grown a lot in the past couple of years. The bigger it gets the more it might change, although not dramatically, because it's never likely to be a million-member service. Not without a radical transformation and a million bucks, anyway.

Three of Echo's many conferences (which are the same thing as forums) are devoted to women's concerns, although the one sponsored by *Ms.* magazine is public and open to both men and women who support feminist ideals. WIT, which stands for Women in Technology, has no focus on technology, really. It seems members just liked the acronym. Of the two closed conferences that are open only to women, WIT has evolved into the one for the Baby Boomer generation, with its counterpart, BITCH—for Babes in Their Cyberspace Hangout—home to the younger women. There's also a section called Men on Echo, known as MOE. They're big on acronyms, obviously.

Like the WELL, Echo also sponsors frequent real-world events. There are regular orientation sessions for new members, weekly informal get-togethers at a Manhattan bistro called The Arts Bar, and two regular series: Read Only (a take-off on a computer term), which is a series of readings by authors who are members of Echo, and Virtual Culture, about the arts and electronic media. The latter reflects the emphasis on the arts in New York City and, therefore, on Echo itself. It's no surprise, then, that the conferences Culture, Movies and TV, and Books are the most popular. The Whitney Museum also runs a conference on Echo, and plans to sponsor special events and art exhibits through Echo's Web site.

Unlike the WELL, which served as a model for Echo, if Echo expands nationally other than through the Internet access it already provides, Stacy has said she would more likely create franchises in other cities that would be run locally and have a local focus, as Echo does, rather than try to grow into a truly national network. For now, the only outpost is Echo's Web site at http://www.echonyc.com (see Fig. 20-1), where Echoids are creating their own Web pages linked to that site.

## Women's Wire

Nancy Rhine says she comes from a family of community-oriented people and educators, so it was only natural that she got into online community building

Welcome to Echo. Echo is New York's premiere online service. On Echo, a diverse group

**20-1** *Echo's Web site, where you can also meet some of the Echoids through their personal Web pages linked to this site.*

when she went to work for the WELL in 1989. After a few years on staff there, she became an independent consultant, but remained an active member. She first met Ellen Pack online, and it was Ellen who encouraged her to start a service for women, helped get the initial financing for it, and became her partner.

"I used to be a labor coach, and I think that's where I got a lot of my strong feelings about empowering women," says Nancy. "I wanted to introduce more women to the world of telecommunications so we could make intelligent choices about whether we wanted it or not. Practical people that women are, I thought if we had more things of interest to women, it would attract more women."

After working toward that goal for 2½ years, Nancy and Ellen officially launched Women's Wire in January 1994, and hired Marleen McDaniel as CEO after an infusion of venture capital, about a year later. "The big vision is to become an international clearinghouse of information online for women, and to have an international community of women online," Nancy says. "So I don't want it to have a regional focus or even a U.S. focus." Because it's San Francisco-based, however, that has been much the aura of it, and that and lack of funds hurt their growth during their first year. That's beginning to change, partly because they've created discussion areas for every major region in the U.S. and some in

Canada, and partly because they've had positive publicity nationwide. So as more women try other services for the first time, more try Women's Wire too.

Because they wanted to make it easy to use, they started with a graphical interface, unlike Echo and the WELL, both of which began as text-only services. But because Women's Wire didn't have major corporate money behind it as services like CompuServe and Prodigy do, they licensed the best independent interface they could find (see Fig. 20-2). It's more than adequate, but also has its drawbacks, including the lack of a built-in address book function. The software was originally designed for Macs, and even though there's now a Windows version it's not always as easy for PC users to get started. A built-in benefit, however, is that you can browse and participate in selected Internet newsgroups. Although the program might not rate four stars, their tech support is stellar. They go out of their way to make sure each new member gets up and running. Like Echo, Women's Wire also has a mentorship program as an option for new members.

Although Wire doesn't have a flagrantly feminist focus, it's fair to say it's an activist-oriented network in that there's an emphasis on social and political

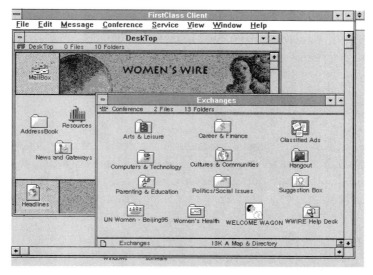

**20-2** *The Exchange, one of the main sections on Women's Wire.*

issues of concern to most women. There's just as much emphasis on the other facets of the lives of women in the '90s too, including careers, children, the arts, health, relationships (in a forum now sponsored by *New Woman*), and personal growth. They've started sponsoring online workshops and plan to add classes, and have introduced monthly themes with guest discussion leaders. You'll find references to some of the specific services throughout this book.

When the Microsoft Network goes online, which is scheduled for August 1995, Women's Wire will sponsor one of two women's forums there as an adjunct to their own separate service and to the one they'll oversee on CompuServe, starting around the same time. It's too soon to tell how these three will differ; Women's Wire itself will continue to be the primary and most in-depth service, yet their partnerships with the other two major services will enable them to offer features and connections, particularly to the Net and the Web, that Women's Wire isn't big enough to provide on its own. Also, some members might prefer to access the more comprehensive service, whereas others will prefer the smaller service especially for women.

Nancy's pragmatic views on the potential impact of online communication on global politics are pretty representative of her views about online advantages in general. "I'm not totally starry-eyed about this," she says. "I just think that if we have people communicating directly, rather than having things filtered through the mass media, that it's got to make a difference, and it will be positive. You get people together and they're concerned about the same things. They're concerned about the quality of their lives, about making a decent living, about taking care of their children and of their parents when they get older. It has the potential of making a major difference. . . .

"I hope that women in general are taught the potential benefits, then they can decide whether they want to be involved with it. There's a way that it has to be taught to people. It's a tool, just like the telephone is a tool. You have your wrench and your hammer, but what if somebody just invented the screwdriver? It's not going to take the place of getting together in person or talking on the phone. People are sensitive to new things and women have been conditioned to think of themselves as not very good at science. There's a mystique about this technology, and that's silly. So I take a lot of time and put a lot of effort and energy into demystifying it. . . . I want people to make informed choices. I would never say to anybody that they should be online if they didn't want to be. You can still have a good life without being online."

## Women's Forum on CompuServe

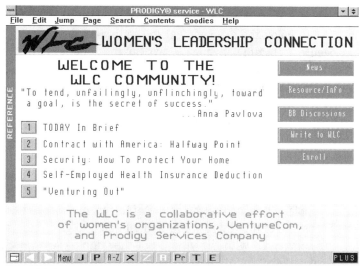

Amy Bernstein, associate editor of *U.S. News & World Report*, deserves credit for getting the magazine to sponsor the Women's Forum on CompuServe, which was the first full-fledged forum (as opposed to only a section of a forum) specifically for women on one of the major commercial services. She and her colleagues staffed it the first year or so in addition to continuing their full-time work as reporters and editors. They feel it deserves full-time staff attention, however, and plan to turn it over to Women's Wire to sponsor later in 1995.

## Women's Leadership Connection on Prodigy

The Women's Leadership Connection (see Fig. 20-3) set up shop on Prodigy in October 1994, and is the brainchild of Susan DeFife and her company,

**20-3** *The Women's Leadership Connection on Prodigy.*

VentureCom, in Washington, D.C. She's the former executive director of Women Executives in State Government, so the thrust of the service is in the direction of political action and business interests, with health in third place. While designing the look and content of the service, she sought input from the National Foundation of Women Businessowners and other business associations, as well as organizations involved with politics and health.

"We wanted to be able to provide good business communication tools, access to the Internet in an easier format, and a private community, all of which we felt was essential to getting women online. It was too expensive to do on our own, but through Prodigy we can provide all that," says Susan. If you want to join Women's Leadership Connection, you must be a member of Prodigy, then fill out an application online and pay an additional $7.50 per month, although access is automatic. Because of the Prodigy affiliation, WLC gained more members in the first six months than Women's Wire added in their entire first year. But the actual participation level has been slower to build, perhaps because it's staffed by only Susan and one employee and they're less experienced as sysops than the leaders on related services, and perhaps partly because of the extra charge.

Over the next year, she plans to recruit new members by going after leaders of key associations and assisting them in getting their members online, which will also enable her to guide the development in the desired direction. "The marketing has been directed toward men," says Susan. "The services have said, 'We've got all these services,' and thrown them out for people to explore. That's what men look for, but women look for ways to build relationships, to expand their networks, and to find relevant information that shows them how to do what they already do, but better and more quickly. . . .

"I am waiting for the day when an alliance forms online or there's an online campaign that effects some change in this country [the U.S.]. I am certain that women are going to be the first to do that, because women want to achieve results. Then people will recognize the power of this technology and the power of women. That's one of the things we're trying to do—to give these organizations a vision. They agree with the vision, but they don't know how to do it yet, so we're actually sitting down and helping them create a plan, starting with e-mail and working up to developing an alliance. It's a cultural change for most of them. It's so different from what they're doing now that we just need to show them how to do it. It's going to take only one alliance or campaign. The first time it happens everybody will want to follow."

# The WELL (Whole Earth 'Lectronic Link)

The WELL, too, is based in the San Francisco Bay area and has been an institution unto itself since it was started by the *Whole Earth Catalog* people. Like Echo, it has always been a text-only service, but they're going graphic in the fall of 1995, by which time they expect to have local-access phone numbers for 80 cities within the U.S. Many WELLbeings, as members call one another, already telnet in from places around the world, via the Internet. See Fig. 20-4 for a glimpse of the WELL through the Web. It's open to visitors at http://www.well.com. For members, it also provides a telnet gateway into the service. The WELL also hosts one of the oldest, if not *the* oldest online sections just for women, called WOW, for Women on the WELL. Although Reva Basch wasn't the founder, she has run the WOW conference for nearly a decade.

The Whole Earth 'Lectronic Link

*Providing Access to People and Ideas*

"A rebirth of community"
- John Markoff, New York Times

"The areas of expertise are astonishing in their diversity"
- Jon Carroll, San Francisco Chronicle

**20-4** *The WELL's well.com mat for visitors. Many members have Web pages of their own linked here.*

"It's eclectic, generally supportive, and a very lively place," Reva says. "People say things like 'Come over here and sit beside me on the sofa' and 'Pass the cookies or wine,' because it feels like a real place. That's silly, but it helps create a mental map of WOW as a physical space. A lot of people think it's about feminist issues only, so they're not interested in joining. We talk about goddess stuff and herbal healing, and feminist and lesbian-feminist politics, but women also talk about the new pair of fabulous earrings they bought yesterday, or log in from Bangkok and give reports on day-to-day life there. So it goes from the trivial to some truly extraordinary reports. . .."

"For a lot of women, online conferencing is very new. They go out into the public conferences and are ignored just because of the pace of the conversation, or because they're new and nobody recognizes them yet, and they feel a little overwhelmed. WOW is a more supportive place to start out. It's also just nice to have a place where there aren't guys. Conversation does take on a different flavor with men involved. They have no problem holding their own in the rest of the WELL, but in WOW you can vent and say 'Why are guys like that?' without having a million people jump down your throat for it. So it's a starting point, it's a refuge, and it's a valuable resource in itself."

Some of the younger women felt out of their element in WOW, so in May 1994 two of them started FemX, where members tend to be under 30, as the name implies. Both WOW and FemX are open to women only, by request, and they also call each and every member at least once to make sure she's a she, rather than some gate-crashing male posing as female.

Marion Davis, of New York City, who cohosts FemX, says she wanted something like FemX rather than WOW because "The women's forums I've seen have turned me off completely. Women need a corner of their own to get to feel more comfortable with a service. But in most of them the attitude is, 'You're complaining about a *date*, while I'm complaining about my child being on drugs!' I felt like a little kid at her mother's tea party.

"FemX members are mostly single, so if the talk is about a mother-daughter relationship they're usually talking about the mother's perspective in WOW and we're more concerned about how the daughter feels. Or because we have a heavy contingent of young journalists, it might be, 'My fourth book just came out, yet I have to temp again [just to cover bills, despite the credentials].' It's a pretty serious conference. A lot of young women are out there trying to make a living any way they can, so they might go online to try to find work. For some reason, it seems to be easier to make women friends online. We get to talk a little more, just sharing our experiences more than you would at a cocktail party—more personal and more open. So it's become a source for a lot of women who are isolated to find friends who are a little more like them."

## Wired Women, on America Online

The newest service on the commercial nets, as of late summer 1995, is Wired Women on America Online. It's a nonprofit network created to coach women in the U.S. on software, as well as how to use computers, get online, and get around online. It will also offer career coaching and perhaps even placement or job referral services, says founder Sarah Browne, in keeping with their slogan: "Get Wired. Get Hired."

## The Internet

The three main newsgroups for women on the Net are soc.women, alt.feminism, and soc.feminism. Of those three, only soc.feminism is moderated, meaning someone is in charge of keeping discussions on track and keeping the flames under control, rather than being a free-for-all. The soc.women newsgroup, according to a study by Ellen Balka, is dominated by men. Only 27 percent of the participants had female-sounding names at the time of her study, although she concedes that it's difficult to tell who's using an alias and about ten percent of the names were gender-neutral. And most newsgroups in the alt hierarchy tend to be a bit, well, you know: alt. Out there. That's their nature and intention. The point is that most women are pretty turned off by the flaming and haranguing that goes on in two of those three newsgroups, and prefer to stick to specialized mailing lists if they use the Internet. There are many of these, with a partial list in the final Highlights section of this chapter.

Speaking of ends, we've nearly come to the end of this book, but I sincerely hope this will be just the beginning of a wonderful new adventure for you. When you get comfortable online, help one more woman along too, please. Feel free to send me e-mail at 74774.1740@compuserve.com or broadhurst@aol.com and tell me what you think of the online world, what you're learning, and any ideas you have for the next edition of this book. Meanwhile, I leave you with this thought from Aliza Sherman of CG Internet Media Marketing, creator of the Webgrrls and SafetyNet Domestic Violence Resources sites on the Web (posted on the Spiderwoman mailing list):

```
To me, "grrl" is a powerful way of saying "girl" and I, at 30, consider
myself to be a girl and grrl (and yes, a woman, too). I think it's a
wonderful, playful, and uplifting reference. . . . I love poking fun at
the staid, dry, stuffed-shirts business attitude, too.

"I love the Web. I feel like I've found a powerful medium that helps me
realize some of my dreams—of publishing, of sharing valuable information
with people, of having fun and making people laugh, of making
connections with interesting people, especially women, all around the
world.

"I spent nine years in the music biz trying to be taken seriously as a
woman (I'm a woman, dammit). But now that I have my own business, I'm
making my own rules, having the time cf my life, and finding more
happiness and success as a Webgrrl than I ever expected.

(Music swells, camera pulls back for wide shot, fade to black. The End)
```

– 330 –

# Highlights: Places online especially for women

The following is an extensive, though by no means complete, list of places online for women:

## Forums or conferences on commercial services

*America Online* Women's Center (also known as Women's Board) —> Exchange —> Community Center —> Women's Center. Also, Wired Women in summer 1995.

*CompuServe* The Women's Forum, started by *U.S. News & World Report* magazine, now sponsored by Women's Wire. Also sections in the following forums: Health and Fitness, Holistic Health, Human Sexuality and

Relationships (HSX200, closed), Issues, Time Warner's Lifestyle Forum, and the Women Online section in ZiffNet's Executives Forum. The League of Women Voters also has a section in the Political Debate Forum.

*Delphi* Domestic/Family Violence Issues, custom forum 111; Locate Deadbeat Parents, custom forum 249; The Mommy Track, custom forum 148; Statuesque and Rubenesque, custom forum 033; The Women's Network, by application to Custom Forum 160; and Women Today, custom forum 386.

*Echo* BITCH (Babes in Their Cyberspace Hangout), for women and by invitation only; *Ms.* Magazine Conference, open to both men and women who support feminist issues; and WIT (Women in Technology), a general-topic discussion forum for women only, by application.

*eWorld* As of early 1995, only 20 percent of eWorld's members were women, which concerned the management, so in June 1995 Becky Boone, the psychologist who also started and leads the Transformations self-help forum, launched another eWorld forum called Worldwide Women Online. eWorld will also sponsor one of the major online feeds from the UN's Fourth International Women's Conference in Beijing, China, in September 1995. As this book went to press, the women's forum on eWorld was not yet open, but Becky says it will include news of legislation and political issues affecting women; sections on the environment, health, lifestyles, recreation and sports, parenting, single mothers, lesbians and bisexuals, women and the arts media, education and careers; and informatin about women's resources on the Internet, as well as Internet courses and workshops. Also an Income of Her Own, a teen-business mentoring program within the Forums section of the Education Center.

– 331 –

*GEnie* The Women Only section of the Family Personal Growth Roundtable and the Women in Management section of the Careers/Workplace Roundtable are open to women only, with access granted by the sysop on request.

*Microsoft Network* Fall 1995, Women's Forum and Women's Wire.

*Prodigy* Women's Leadership Connection and the Women's Issues section in the Lifestyle Bulletin Board. There's also a women's section within the Stocks section of the Business/Finance Bulletin Board.

*The WELL*  Women on the Well (WOW) and FemX, both by application and open to women only.

*Women's Wire*  All of Women's Wire is mainly for women, but they also have the following special conferences: Older Women's Network, Women's Spirituality, Charlitas (for Latinas), and Sappho & Friends, all in the Cultures & Communities section; New Media Women, in the Career & Finance area; and Gender, Race, & Class, in the Politics and Social Issues section. Also see the earlier sections, *CompuServe* and *Microsoft Network.*

# World Wide Web sites

Disclaimer: Internet addresses can change or be cancelled without notice, and new things are added daily. Please see appendix B about what you need to use the Web.

*The Clearinghouse for Subject-Oriented Resource Guides*  By the University of Michigan's Library and School of Information and Library Science. Excellent women's studies and women's health info and leads. Web URLs are:

http://www.lib.umich.edu/chhome.html
http://http2.sils.umich.edu/~lou/chhome.html
http://una.hh.lib.uich.edu/11/inetdirs

*Feminist Activist Resources on the Net*  By Sarah Stapleton-Gray. The Web address is: http://www.clark.net/pub/s-gray/feminist.html. Features an index including Communicating with Other Feminists, Current Feminist Issues: News and Resources, Reproductive Rights, Sexual Harassment and Rape, Domestic Violence, Women of Color, Women and Work, Other Topics, Women's Organizations, Feminist Resources, General Resources for Political Activists, Feminist Activist Calendar, Take Action: Suggestion for Current Feminist Action, and Feminist Fun and Games.

*Feminism and Gender Issues*  By Laura Hunt. Web: http://tiger.cc.uic.edu /~lauramd/fem.html

*Gender resources*  Web: http://english-server.hss.cmu.edu/Gender.html

*Global Fund for Women*  Web: http://www.ai.mit.edu/people/ellens /gfw.html

*Men's issues, M.E.N. Magazine*  Web: http://www.vix.com/menmag/

*Self-defense, model mugging program*  Web: http://www.ugcs.caltech .edu/~rachel/bamm.html

*Women's Home Page*  By Jessie Stickgold-Sarah. The Web address is: http:// www.mit.edu:8001/people/sorokin/women/index.html

*Women and Computer Science*  Web: http://www.ai.mit.edu/people /ellens/gender.html

*The Women's Page of the San Francisco Examiner*  Web: http: //sfgate.com/examiner/womenweb.html

*The Women's Resource Project*  Web: http://sunsite.unc.edu /cheryb/women/wshome.html

*The Women's Resources Project Resource Page*  Web: http://sunsite.unc .edu/cheryb/women/wresources.html

*Women's Studies Database*  Web: http://info.umd.edu:86/Educational _Resources/AcademicResourcesByTopic/WomensStudies–> inforM

# Internet mailing lists

*Breast-Cancer-L*  E-mail to: listserver@morgan.ucs.mun.ca
Message: subscribe breast-cancer

*CFS-L (Chronic Fatigue Syndrome list)*
E-mail to: listserv@list.nih.gov
Message: subscribe cfs-l (your first and last name, no parentheses)

**CFS-Newsletter**  E-mail to: listserv@list.nih.gov
Message: sub cfs-news (your first and last name, no parentheses)

**CFS-Wire (Chronic Fatigue Syndrome newswire)**
E-mail to: listserv@stjohns.edu
Message: subscribe cfs-wire (your first and last name, no parentheses)

**CITNET-W (Healthy Cities Women's Network)**
E-mail to: listserv@indycms.iupui.edu
Message: subscribe citnet-w (your first and last name, no parentheses)

**Depression, experience, and treatment**
E-mail to: listserv@soundprint.brandywine.american.edu
Message: subscribe depress

**Diabetes**  E-mail to: listserv@lehigh.edu
Message: subscribe diabetic (your first and last name, no parentheses)

**Diet and diet discussions**  E-mail to: listserv@ubvm.cc.buffalo.edu
Message: subscribe diet (your first and last name, no parentheses)

**DDFind-L (disability networking)**  E-mail to: listserv@vm1.nodak.edu
Message: subscribe ddfind-l (your first and last name, no parentheses)

**WitsEnd (endometriosis)**  E-mail to: listserv@dartcms1.dartmouth.edu
Message: subscribe witsend (your first and last name, no parentheses)

**FAST (fight against sexist tyranny)**  E-mail to: listserv@gitvm1.gatech.edu
Message: subscribe fast (your first and last name, no parentheses)

**FeMail**  E-mail to: femail-request@lucerne.eng.sun.com
Message: subscribe femail (your first and last name and gender, no parentheses)

**Feminism digest (digest of soc.feminism newgroup)**
E-mail to: feminism-digest@ncar.ucar.edu

Message: subscribe feminism-digest (your first and last name, no parentheses)
Post responses to messages to: feminism@ncar.ucar.edu

*FEMISA (on feminism, gender, women, and international relations)*
E-mail to: listserv@csf.colorado.edu
Message: subscribe femisa (your first and last name, no parentheses)

*Femrel-L (feminism, theology, and religion)*
E-mail to: listsrv@mizzou1.missouri.edu
Message: sub femrel-l (your first and last name, no parentheses)

*Fit-L (fitness, diet, exercise)*  E-mail to: listserv@etsuadmn.etsu.edu
Message: subscribe fit-l (your first and last name, no parentheses)

*Study of communication and gender*  E-mail to: comserve@vm.its.rpi.edu
Message: subscribe gender (your first and last name, no parentheses)

*General list on gender issues*  E-mail to: ericg@indiana.edu
Subject line: add me mail.gender
Message: leave blank

*Granite (general list on gender issues)*  E-mail to: listserv@nic.surfnet.nl
Message: sub granite (your first and last name, no parentheses)

*IHP-NET (interfaith health practices)*
E-mail to: majordomo@interaccess.com
Message: subscribe ihp-net (your e-mail address, no parentheses)

*KOL-ISHA (a moderated list about women's roles in Judaism)*
E-mail to: listserv@israel.nysernet.org
Message: subscribe kil-isha (your first and last name, no parentheses)

*MAIL-MEN (gender issues from a male perspective)*
E-mail to: mail-men-request@usl.com
Message: Make inquiry to a live person about joining the list.

*Mensig (men's special-interest group)*
E-mail to: listserv@mizzou1.missouri.edu
Message: subscribe mensig-L (your first and last name, no parentheses)

*Pro-Feminist Men's Issues Mailing List*
E-mail to: jyanowitz@hamp.hampshire.edu
Message: Make inquiry to a live person about joining the list.

*RURWMN-L (rural women)* E-mail to: listserv@bingvmb.cc.binghamton.edu
Message: subscribe rurwmn-L (your first name and last name)

*SASH (sociologists against sexual harassment, moderated)*
E-mail to: azpxs@asuvm.inre.asu.edu
Message: Make inquiry to a live person about joining the list.

*South Asian Women's Net*
E-mail to: usubrama@magnus.acs.ohio-state.edu or susanc@helix.nih.gov
Message: Make inquiry to a live person about joining the list.

*SWIP-L (Members of Society for Women in Philosophy and others, on feminist philosophy)* E-mail to: listserv@cfrvm.cfr.usf.edu
Message: subscribe swip-l (your first and last name, no parentheses)

*StopRape (a sexual assault activist list)*
E-mail to: listserv@brownvm.brown.edu
Message: subscribe stoprape (your first and last name, no parentheses)

*Wellness (health, wellness, nutrition, and life expectancy)*
E-mail to: majordomo@wellnessmart.com
Message: subscribe wellness

*WHAM (Women's Health Action and Mobilization, for those who agree that "women's health is political")* E-mail to: listproc@listproc.net
Message: subscribe wham (your first and last name, no parentheses)

**WIG-L (sponsored by the Coalition of Women in German)**
E-mail to: listserv@cmsa.berkeley.edu
Message: sub wig-l (your first and last name, no parentheses)

**Women** E-mail to: women-request@athena.mit.edu
Message: subscribe women (your first and last name, no parentheses)

**Women (domestic violence, harassment, education, sexual diversity, health, and child and parent care)** E-mail to: majordomo@world.std.com
Message: subscribe women (your first and last name, no parentheses)

**WMN-HLTH (Women's Health Electronic News Line, sponsored by the Center for Women's Health Research)**
E-mail to: listserv@uwavm.u.washington.edu
Message: subscribe wmn-hlth (your first and last name, no parentheses)

**WMSPRT-L ("For women and men interested in goddess spirituality, feminism, and the incorporation of the feminine/feminist idea in the study and worship of the divine.")** E-mail to: listserv@ubvm.cc.buffalo.edu
Message: subscribe wmsprt-l (your first and last name, no parentheses)
To post messages: wmsprt-l@ubvm.cc.buffalo.edu

**WMST-L (on women's studies)** E-mail to: listserv@umdd.umd.edu
Message: subscribe wmst-l (your first and last name, no parentheses)

**WON (the Women's Online Network; political activism)**
E-mail to join: carmela@echo.panix.com or horn@echo.panix.com
Phone: 212-255-3839 in New York City
$20 per year subscription fee, negotiable for those in financial difficulty

**YEAST-L (yeast-infection-related medical issues)**
E-mail to: listserv@psuhmc.hmc.psu.edu
Message: subscribe yeast-l (your first and last name, no parentheses)

# Internet newsgroups (Usenet)

Software for commercial services, such as America Online and CompuServe, as well as many Internet programs such as Netcruiser and Pipeline, enable you to join newsgroups automatically. You can also just browse most of them on Delphi and selected ones on Women's Wire.

alt.abortion
alt.abortion.inequity
alt.abuse-offender.recovery
alt.abuse.recovery
alt.abuse.transcendence
alt.adoption
alt.adoption.agency
alt.child-support
alt.discrimination
alt.feminism
alt.infertility
alt.missing.kids
alt.parents-teens
alt.psychology.personality
alt.recovery
alt.recovery.addiction.sexual
alt.recovery.codependency
alt.sexual.abuse.recovery
alt.society (many)
alt.support
alt.support.abuse-partners
alt.support.anxiety-panic
alt.support.arthritis
alt.support.asthma
alt.support.attn-deficit
alt.support.big-folks

alt.support.cancer
alt.support.cerebral-palsy
alt.support.crohns-colitis
alt.support.depression
alt.support.dev-delays
alt.support.diabetes.kids
alt.support.diet
alt.support.dissociation
alt.support.divorce
alt.support.eating-disord
alt.support.epilepsy
alt.support.headaches.migraine
alt.support.loneliness
alt.support.mult-sclerosis
alt.support.non-smokers
alt.support.obesity
alt.support.short
alt.support.shyness
alt.support.step-parents
alt.support.stop-smoking
hiv.actup-actnow
hiv.aids.issues
hiv.alt-treatments
hiv.announce
hiv.med.questions
hiv.planning

hiv.resources.addresses
misc.activism.progressive
misc.consumers
misc.education
misc.education.adult
misc.education.home-school.christian
misc.education.home-school.misc
misc.education.learning-disab
misc.health.aids
misc.health.alternative
misc.health.arthritis
misc.health.diabetes
misc.jobs.contract
misc.jobs.misc
misc.jobs.offered.entry
misc.kids
misc.kids.health
misc.kids.info
misc.kids.pregnancy

sci.life.extension (slowing aging)
sci.med.dentistry
sci.med.diseases.cancer
sci.med.nursing
sci.med.nutrition
sci.psychology
sci.psychology.digest
sci.psychology.research
soc.bi
soc.couples
soc.couples.intercultural
soc.culture (many)
soc.feminism (moderated)
soc.gender-issues
soc.men
soc.women
talk.abortion
talk.rape

# Internet gophers

*Abortion and Reproductive Rights* //gopher.well.sf.ca.us/11s/politics /abortion

*The Clearinghouse for Subject-Oriented Resource Guides* By the University of Michigan's Library and School of Information and Library Science. Excellent women's studies and women's health information and leads. //gopher.lib.umich.edu

*Institute for Global Communications* Excellent and extensive list of e-mail addresses for women's organizations worldwide. //gopher.igc.apc.org:70 /00/women/Directory/directory

*National Center for Nonprofit Boards* //ncnb.org:7002/1

**Women's Studies and Resources** //gopher.peg.cwis.uci.edu:7000/11 /women or //gopher.peg.cwis.uci.edu 7000

**The Women's Wire Gopher** //gopher.path.net:8101/ or //gopher.path.net 8101

# The Institute for Global Communications' WomensNet

You must belong to the IGC online network to belong to WomensNet. For more information, send e-mail to womensdesk@igc.apc.org or women-info@igc.apc.org. The following are online conferences on WomensNet:

*amlat.mujeres* For women in Latin America and the Caribbean, with a focus on regional problems and concerns.

*apngowid.meet* The proceedings of APNGOWID (Asia-Pacific NGO Women in Development); A November 16–20, 1995 meeting in Manila.

*econ.women* To foster ideas and actions towards the creation of an economic system based on other than monetary values. We are particularly interested in women's perspectives and welcome ideas from women that further this goal.

*hivnet.women* Issues confronting women with arc/hiv/aids. Sponsored by:

HIVNet, HIV Nereniging Nederland
Postbus 15847, 1001 NH
Amsterdam, The Netherlands
phone 31-20-664-4076, fax 31-20-664-6689, e-mail hivnet@ooc.uva.nl

*hr.women* Information and materials on human rights issues specific to women.

*icpd.general* General discussion and the exchange of documentation for the 1994 International Conference on Population and Development.

*lambdaletters* Monthly sample letters prepared by the Lambda Letters Project to help people write to legislators and other public officials about women's issues, lesbian/gay/bi issues, people of color issues, and HIV/AIDS.

*lescon* Lesbian Contradiction: A Journal of Irreverent Feminism is a quarterly of opinion, testimony, cartoons, and humor by and for women who aren't afraid to disagree or challenge each other.

*muj.beij95* September 1995 International Women's Conference in Bejiing, Spanish version.

*population* Forum to discuss the causes and possible solutions to the problem of overpopulation.

*ppp.meet* Proceedings from the People's Perspectives on Population International Symposium, Bangladesh, (12–15, December 1993) as a preparation for ICPD 1994 conference.

*un.wcw.doc.eng* Official UN documents for the Fourth World Conference on Women, in English.

*un.wcw.doc.esp* Official UN documents for the Fourth World Conference on Women, in Spanish.

*un.wcw.doc.fra* Official UN documents for the Fourth World Conference on Women, in French.

*wcw.ngo.doc* NGO documents related to the World Conference on Women and NGO Forum.

*wfs.english* Articles from Women's Feature Service (WFS), based in New York.

*wilpf.hotline* Regular updates on legislation on such issues as disarmament, racial justice, and an end to U.S. global intervention. Sponsored by the Women's International League for Peace and Freedom.

*women.comms* A place for discussion and resources of all forms of media in use in many areas of work.

*women.dev* Information relating to local/regional/international development issues affecting/involving women.

*women.east-west* A forum to discuss gender issues in the radically altered societies of East and Central Europe and the former Soviet Union. Sponsored by the Network of East-West Women (NEWW).

*women.events* Information related to events/programs/courses/conferences organized by or for women.

*women.forum* Discussion of women's issues, open to all.

*women.health* Information relating to women's health and related issues.

*women.info.src* This moderated conference contains detailed listings of information sources pertaining to women's's issues: databases, live or electronic networks, books, clearinghouses, documents, libraries.

*women.labr* News/information/discussion concerned with women and labor issues.

*women.news* News items and announcements regarding women and women's issues.

*women.only* A private discussion place for women only.

*women.population* Resource information about population, specifically as it relates to women.

***women.unwcw***  For women's organizations and information providers.

***women.violence***  News, information, and discussion relating to all acts of violence against women.

And these are women's organizations on WomensNet:

Akina Mama Wa Africa
All China Women's Federation
American International Health Alliance
Brigham and Women's Hospital
American Women's Expedition
Border Women's Communication Network
Boston Women's Health Book Collective
Casa de Colores
Center for Women's Global Leadership
Centro de Investigacion y Capacitacion de la Mujer
Change of Heart Chicago
Foundation For Women
DSA Feminist Commission
Equality Now
FEMNET
Foundation for a Compassionate Society
Global Fund for Women
Grupo de Saude da Mulher
International Women's Health Coalition
International Women's Tribune Centre
ISIS International
League of Women Voters
Mujer a Mujer
National Women's Network
OISE-Centre for Women's Studies in Education
Women's Foreign Policy Council
Women's Environmental Network
Women's Feature Service (also available elsewhere)
Women's International League for Peace & Freedom
Women's International News Gathering Service

Profuse thanks to the following people for their help or permission to use part of their materials in compiling the previous lists:

➤ Angela Gunn, editor of *Web Week* and author of *Plug-n-Play Mosaic and Web Guide: Exploring the Weird, the Wild, and the Wonderful on the World Wide Web* (see the Bibliography)

➤ Munn Heydorn of The First National Bank of Chicago, who maintains the Internet Resources for Not-for-Profits in Housing and Human Services directory, which is available on the Web at http://www.ai.mit.edu/people /ellens/Non/online.html

➤ Laura Hunt, who compiled and maintains the Sources for Women's Studies/Feminist Information on the Internet index

➤ Sue Mooney of WomensNet, sponsored by the Institute for Global Communications

➤ Nancy Rhine and Ellen Pack of Women's Wire and their list, Electronic Forums of Interest to Women

➤ Sarah Stapleton-Gray, who compiled and maintains Feminist Activist Resources on the Net, on the Web at http://www.clark.net/pub/s-gray/feminist.html

# Ⓐ
# Online tools

## You can get there
## from almost anywhere

Even if you're still using an IBM XT with only floppy disk drive or an early Mac, you can get online. Unless you have a faster computer, a good amount of RAM, and a sizable hard disk, however, you might not have quite the "Wow!" reaction that most people do when they first log on—feeling they've suddenly catapulted into another dimension. Still, you'll probably agree with what Paula Span said after her first weeks on The WELL, which at the time had no flashy graphics at all: "It all looks quite prosaic at this point—just lines of text appearing on my screen—but it feels very exciting."

With an older computer, you can also access the Internet, CompuServe, Delphi, and GEnie (until they change things, of course); you just won't get the full effect, with photos and graphics. You don't need an IBM-compatible or a Mac, either. Amigas, Commodores, and other kinds of computers will work for some services. Just make sure you ask when you sign up whether your computer is supported on that network.

Of course, you can't get online without a modem. That's the key component to transforming your computer from a mere machine into a powerful communication tool. The modem connects your computer to the phone line and lets the phone line connect you to the computers that run the online services. Nearly all cities and many smaller towns now have local, toll-free numbers that connect to the major services, as well as local BBSs (bulletin board systems) and Internet-access providers in many areas. Some offer low flat fees for monthly access to an 800 number if they don't have a local-access number.

# All you need to know about modems

Forget all that cryptic jargon about baud rates and IRQ settings that has intimidated you and kept you from doing all the terrific things you can do once you have a modem. It's actually very simple.

## The two types of modems for desktop computers

All you need to know about modems for desktop computers is that they come in two basic types (see Figs. A-1 and A-2) and several different speeds, and that the type doesn't make much difference but the speed does. The two kinds are:

*Internal modems* These plug into a slot inside your computer, anchored by one screw at the back. They're sometimes called *cards* or *boards* because they're circuit boards. They cost less and take up no desk space, but can be trickier to install depending on how many other extras your computer has. Many new computers come with modems already installed, so you might not have to bother with this at all.

*External modems* This kind plugs into a port at the back of the computer. The external modems cost a bit more, but they're easier to install as long as you have an available serial port. The flashing lights also reassure you that they're working, and come in handy for troubleshooting.

**A-1**  *Internal modem, from U.S. Robotics.*

**A-2**  *External modem, from U.S. Robotics.*

# The two types of modems
# for portable computers

If you use a laptop or a notebook computer, you also have two choices:

*PC-card modems*  Card modems were known as PCMCIAs, so you might still hear people call them that until everybody gets used to the new name. No one could ever remember what all those letters stood for, so the industry jargon arbiters dropped the term, thankfully. These are about the size of a credit card and slide into a slot in your computer (see Fig. A-3). As they get more refined, they'll eventually be used in desktop systems. They cost considerably more than detachable modems, but use less of your working memory (RAM) and help batteries last longer.

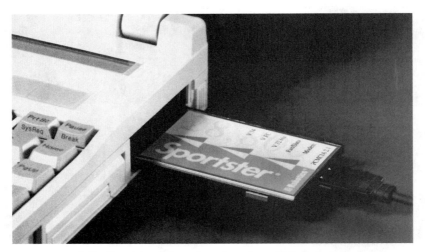

**A-3** *Sportster PC-card modem from U.S. Robotics.*

***Portable modems*** This kind attaches to the computer when you want to use the modem. It works more easily with some software than the card modems, but they'll soon be about equal in performance, which will make a portable modem more of a hassle while traveling because it's heavier and takes up more space. For now at least this kind costs less, although both cost more than the desktop modems do, just as portable computers cost more than the ones that stay put.

## Speed, the crucial factor

All modems work basically the same, and all must be connected to phone lines. It's what happens when they're connected that makes the difference, and that depends on how fast they can transmit data (send and receive information between computers or between your computer and an online network).

Modems come in several speeds, but bigger numbers are better because bigger means faster. The faster they move data, the less time you spend connected, thus the more money and time you save. You'll seldom see any modems in computer stores or catalogs slower than 14.4 or 28.8 Kbps (thousand bits per second). Even if someone offers to give you a 2,400-bps modem, tell 'em thanks, but no thanks, unless you intend to stick with text-only services and e-mail messages, and won't be sending or receiving large document files or software programs. Don't buy a 9,600-bps modem either; a 14.4 or 28.8 is a far better investment. Besides, you'll pay as much for a 9,600-bps as a 14.4-Kbps modem.

*Note:* If you have a PC that's older than a 386, to use a high-speed external modem (faster than 9,600 bps) you'll also probably need to upgrade what's called the uart, at which point it's time to talk to a tech-savvy friend, unless you're sophisticated about such matters yourself, in which case you won't likely be reading this section anyway. For internal modems, shop for one with a 16650A uart chip on the card. Either way, make sure the uart is a function of the hardware, not software.

As with most purchases, you get what you pay for in modems. Stick with a major name brand, such as U.S. Robotics, Hayes, or Practical Peripherals, and any 14.4- or 28.8-Kbps modem will probably be fine. Really.

## Other relevant modem specifications

If you really want all the details, read this section. If not, don't worry about it.

Look for a 28.8-Kbps modem that operates on the v.34 protocol, which is the new international standard. For 14.4 Kbps, v.32 is the standard. Don't let all these esoteric numbers confuse you; just jot them down on your shopping list and look for a modem that matches the appropriate specs. Here's what the experts recommend: v.42 error correction, v.42 bis compression, MNP 4 and MNP 5, and for fax standards group III, class 1, and class 2. Also, make sure that all this is built entirely into the hardware, rather than working through software.

Virtually all modems now on the market come with software to make them work, although sometimes it's not very good or simply a scaled-down version of a popular program, so you'd be better off buying separate software. Most modems can also send faxes, although don't take that for granted. Ask. Again, quality varies, so make sure the specs match the specifications in the previous paragraph.

And yes, the same modems work for both PCs and Macintoshes if they'er external. They use different software and Macs need a cable made for that system, but most will come bundled with whatever you'll need.

*Caution:* If you install your own internal modem, make sure you ground yourself by touching metal—such as the computer case—before touching anything inside your computer, so you don't short something out due to static electricity. Hold the modem card by touching only the edges or blank spaces, as you would a purse-size mirror or photograph. Never touch the little gold-colored contact points at the bottom of an internal modem card or you could fry it. Take it from one who learned the hard way—this is not a new culinary technique.

Call the tech support line for the modem manufacturer if you have trouble installing it, but a good instruction brochure will usually get you through the process in a few minutes. If a capable friend offers to install the modem for you, by all means graciously accept.

To recap, it comes down to this: All you really need to know is whether you want an internal or external modem, then what's the fastest one you can afford. Simple.

# Software you need to get online

Because the modem is simply hardware, it just sits there, useless, without software to tell it what to do. Many modems come with free plug-and-play software for major online services. Easy. But they work only for those services. To do other things you'll need multipurpose communications software, or what's casually called a *comm program*. The rule for buying software is always to decide what you want to do, then get the best tool for that job. This means you'll need two basic kinds of software to try a variety of online services (don't worry, because it's not as complicated as it sounds):

# All-purpose communications software

A communications program that works with almost any service gives you access to things like BBSs, freenets, and basic, text-only Internet. It can even get you onto some major commercial services such as CompuServe, assuming you're willing to find your way around manually by typing all of the commands rather than taking advantage of their automated software.

It's better to use an automated program whenever you can, but there are times you'll need a generic communications program. That's what you use to send something to anyone who has a computer with a modem but who isn't online, or maybe your office if you work elsewhere but are working at home that day.

You can start with the software that comes with your modem (or what is integrated into Windows or the Mac operating system) if you intend to use only BBSs, freenets, or commercial services that provide their own software. With a few exceptions, such as CrossTalk Communicator, it's usually either a scaled-down version of a fancier program and they're hoping you'll pay to upgrade to the full program, or else it's software that's not as easy to use as the pricier programs and rather limited in what it can do.

So it's a good idea to upgrade to a program that's easier to use and more sophisticated than what usually comes with the modem.

For communications software that's easier to use and more flexible, Delrina's WinComm, Procomm Plus, and CrossTalk are the most popular for PCs, with WinComm, Microphone, and White Knight often the choice of Mac users. There are also excellent shareware and small-vendor programs available for communications only (rather than combination communication and fax) software, such as Telix for DOS or Z-Term for the Mac. Remember, you can use what your modem comes with until you're more sure of what you want to do.

A note about fax software: Even if the prepackaged software is capable of sending and receiving faxes as well as making ordinary modem connections, most of it will transmit only text and won't handle graphics or even fancy type on letter-heads, no matter what the blurbs on the boxes say. For graphics, you need soft-ware with OCR (optical character recognition) capabilities, such as WinFax Pro or Eclipse Fax, and at least 8MB of RAM. However, for straight text you can use the basic software that comes with the modem. You can also send straight-text faxes through some of the online services (for a fee, of course).

# Proprietary software for specific services

Depending on what services you want to try, you might need specialized or pro-prietary software for access to certain services, such as CompuServe's optional WinCIM, or America Online's or Prodigy's programs, which are the only way you can access those services. Generically, these are known as *front-end* or *offline reader* programs, depending on how much of the action they automate. If they have graphics that can be activated by a mouse or trackball, or point-and-click icons, they're also called GUI (pronounced GOOey), meaning graphical user interfaces. They automate complicated commands and make it all so easy that all you have to do is choose what you want to do by pointing to pictures.

Some of these programs enable you to read and write messages or even browse forum messages and libraries offline, without the meter running. Of course you have to do a go-and-grab session first to download stuff to choose from, but that can take as little as two or three minutes. You can hit one or two keys, go make your morning coffee or walk the dog, and come back to find your e-mail waiting and updates on discussions you were following in your favorite forums. Start with whatever the online service recommends, even if it's not the fastest and most flexible and powerful. It will be the easiest, and you can switch to an alternative program once you know your way around (see box for suggestions). See the following for suggestions:

> If you have an older-model, DOS-based computer, here are some programs that take up far less disk space, yet automate commands to make things easier and faster and save you money (they're all available to download once you get online):

**CompuServe** Tapcis, which is an excellent, powerful, flexible program (phone number in the U.S. is 1-800-SUPPORT, or GO Tapcis online), and Autosig (GO PCNet) and OzCIS for DOS (GO OzCIS). There's also a DOS version of CIM, the CompuServe Information Manager, although it's not as slick and intuitive as the WinCIM version (GO DOSCIM).

**Delphi** D-Lite, in the Computing Forum (–>D-Lite Support), or Rainbow, also in the Computing Forum.

**GEnie** Aladdin, in the Aladdin Roundtable library.

Once you feel like you know your way around a bit and understand how the system is set up, you can automate all your mail and most of your forum and library visits with CompuServe Navigator, available for Windows and the Mac, and both OzCIS and NavCIS, for Windows—all available from CompuServe. Just type any of these names at the FIND command. There are other programs available for CompuServe and Delphi, so ask in the new-member forums, which are free.

# Computer system must-haves

While it's true that you can get online, somewhere, somehow, with an older-model computer that doesn't even have a hard disk, you won't get very far and you might not get where you really want to go. However, do not, repeat, do *not* let that keep you from getting online. It's typically the people who equate the size of hard drives and RAM speed with sexual prowess who insist that computing is an all-or-nothing proposition. Either that or they sell them and work on commission. I managed just fine as a sysop (systems operator, or online staff) on CompuServe with an XT clone with an amber monitor and a 40-meg hard drive, and sometimes with a laptop with 640K of RAM and no hard drive at all, and still could. So start with whatever computer system you have or have access to, adapt, and move up from there when you can.

But if you want to be able to take advantage of all that's available right away, you'll need more. Much more. All these flashy graphics, photos, and sound bytes that the online services now use demand faster, more powerful computers with much bigger hard drives for data and software storage than what was quite adequate only a year ago. Just one program can take up as much as 40 megabytes on a hard drive, and that's not counting the space you'll need for all the goodies you'll find and want to download. For RAM, which is what your computer uses while it's running programs and working, even eight megs is minimum now for PCs and Macs, but with Windows 95, PCs will run better with 12 megabytes of RAM, preferably 16. Super VGA color monitors are now essential and, because Windows programs need wider screens, a 15-inch screen is minimum.

And unless you want to sit there for two or three hours feeding disks into your computer to do a back-up—which means you'll never get around to it, and really regret that the day something goes wrong that damages or destroys what you've stored—a tape back-up drive is also basic equipment, and a very good investment. Get one. If you're going to have direct access to the Internet, you should also have a virus-check and fix-it program and run it daily. Finally, you need a system-monitoring, maintenance, and repair program, such as Norton Utilities or PC Tools, if you want to keep your anxiety level low and have a first-aid kit in case of emergencies, so get those too.

# Add-ons for the wish list

That's just the must-have shopping list. You might as well order a CD-ROM drive too, because kids (and grown-ups) want them for games and all kinds of reference discs that you suddenly realize you can't live without, such as medical encyclopedias with fully detailed diagrams and dissections of the body. In 256 colors, no less. The online services now offer extras on CDs, too. Any program that can run from a CD saves precious space on your hard drive, but runs significantly slower than those installed on the hard drive. That will improve, though, and more business programs will be released in CD format in the next year or two.

Oh, and you'll need a sound card and speakers for the CD-ROM drive, of course. Even if you put off buying the drive until later, to hear the sound bytes from *Newsweek*'s interactive edition on America Online or the latest from the Underground Music site on the Web, you'll need a SoundBlaster 16 or other good sound card and special speakers. Like all speakers, they range rather widely in quality. The Labtecs won't sound as good as NECs, but they won't cost as much either. As always, you get what you pay for, but if you're using them only for games and not music, it won't make much difference.

All this means that you need at least three drive bays: one for 3.5-inch disks, one for the CD drive, and one for the tape back-up. If you use a PC and still want to use those old 5.25-inch floppies, you'll need four drive bays, so you might as well consider a tower design that sits on the floor rather than a desktop model. With either one, the whole shebang is a full-scale "multimedia computer," which amounts to serious money. But bear this in mind: It's cheaper and safer to buy a new computer system than to rebuild the old one (another lesson learned the hard way).

# Smart shopping tips

Ah, but there is a bit of good news. Sorta. You can often negotiate the price of the total package, especially if you add extras. This is good. You'll always have to add extras, such as more RAM, which is not so good. Here are a few more tips:

You must do your homework first and go into it prepared and on the offense, because vendors have been caught quoting higher prices to women than to men, just as they have with cars. Read all 800 pages of *Computer Shopper* catalog cover-to-cover if you have to in order to learn what's what. *PC/Computing, PC World, ComputerLife, HomePC* and *Home Office Computing* are also reliable magazines, and easier to understand than ones for more advanced users, such as *PC Magazine*. It's no different than shopping for a stove or a microwave oven. If you're going to buy, be an informed buyer.

Discount superstores sometimes have great prices because they're selling close-out models. You can usually get better deals on current models at good mail order dealers, such as Microwarehouse or Mac Warehouse or PC Connection and Mac Connection, Dell, Gateway or Northgate.

People who always have to have the latest, fastest, most powerful computers are unloading 486 and 386 PCs in favor of Pentiums. Thus, there are bargains available, and a used computer could serve you well for a year or two. Check your local newspaper for the local PC users group or call Boston Computer Exchange (800-262-6399 in the U.S.), which guarantees its used computers and serves as the go-between until the buyer and seller are satisfied. But invest the 50 bucks or so that a consultant or local shop charges to run diagnostic tests overnight to make sure there are no problems with the machine before you make a final commitment. A reputable and honest seller will agree to that.

Never buy new operating system software (even the much-ballyhooed Windows 95) until it's in its second or third version, because there are always glitches, also known as bugs. Nobody wants bugs of any kind. So if it says version 1.0, don't even consider it. Even if it says version 3.0, pass. Wait for version 3.1 or 3.2.

Whatever your priorities, follow the standard advice: Always get more than you think you need and budget for the best you can possibly afford. Then expect to spend another $200 or so beyond that.

# Computer buying checklist

☑ The fastest, most powerful computer you can afford, and one that's capable of more than what you think you'll need. Otherwise, you'll surely change your mind and wish you'd bought one level up right after the return period elapses.

☑ For a new machine, get a 486/33 model in a PC, minimum (386/25 minimum for used ones). Higher numbers mean more power and more speed, which means less frustration and more options. For a new Macintosh, make sure it has a System 7 operating system. There's much less difference in difficulty between PCs and Macs than there once was, and if you can use one, you can easily learn the other. So get whichever kind the most people in your household are used to using at work or at school.

☑ 14.4 Kbps or 28.8K modem, internal or external, for desktops; PC card or portable for laptop or notebook computers.

☑ 8 megs RAM minimum for Macs and PCs, preferably 12 or 16 for PCs.

☑ Nothing less than a 540-meg hard drive, up to a gigabyte if you can afford it, for both Macintoshes and PCs.

☑ Anti-virus and system maintenance software, preferably compatible, such as Norton Antivirus and Norton Utilities.

☑ Fax software, preferably with OCR (optical character recognition). If you use Windows, get a Windows fax program too, because it's easier and quicker to send faxes directly from the document you're working on.

☑ Communications software better than what's built into the operating system, although there's no great advantage to fancy icons other than it's slightly easier to use and looks nicer. But it also takes up more room on your hard drive.

☑ Automated-access software for commercial services you intend to use, which is usually free when you sign up for services such as America Online, CompuServe, Delphi, GEnie, Interchange and Prodigy. Software for Microsoft Network is built into Windows 95, or whatever they end up calling the new version.

☑ Automated-access Internet software if you expect to roam the Internet more than five hours a month. Make sure it has a forms-compatible Web browser, such as Internet in a Box, Netcom's Netcruiser or PSI Net's Pipeline. Netscape is also highly regarded, but non-techies may need help installing it, whereas the others install almost automatically. (See Appendix B.)

☑ Tape back-up drive and software to make it work, which is usually bundled in the same package. Colorado Tape Back-up System is one of the most popular, and easy to use.

# B
# Internet tools

Let's say you've checked out America Online, CompuServe, Delphi, and Prodigy and have decided that you've just gotta be on the Net. Sure, you can get there through the commercial services, but if you expect to use the Net a lot you'll save money if you get direct access rather than paying to go through their gateways. The break-even point is at about five hours a month strictly Net time. Less than that, stay with the commercial services; more than that, read on.

The next recommendation is that you ignore anyone who tells you about all the free software that's available and how you can learn all the UNIX commands to navigate the Net raw, and ignore the computer cowboys or techno-jocks who tell you only wimps use graphical interfaces or automated software. Who cares? Unless you have a masochistic streak, wimp is good. There's no reason to do it the hard way when there are so many easier ways available.

Assuming you already have at least a 14.4-Kbps, high-speed modem, you have two fairly simple choices to make before you can take up permanent residence on the Net: You need to choose an Internet service provider and, maybe, which software to use.

# How to choose an Internet service provider

Again, skip the back-to-basics approach by making sure that the provider you choose can give you access to the World Wide Web. That's the part of the Internet that has graphics, photos, and sound. Visually, exploring the Net via plain text, rather than via the Web, is like the difference between watching black-and-white and color TV. In terms of getting where you want to go quickly and easily, navigating by text commands rather than the Web is like the difference between a bike and a car. The Web has things that aren't anywhere else on the Internet, but you can get to anything that's available elsewhere on the Internet as long as you have a good Web browser program that includes other basic Internet software tools to use gopher and transfer files (FTP) and telnet to other locations. Most do.

Without a Web browser, you've narrowed your options considerably. You'll lust after it anyway as soon as you get a glimpse of what's there. The Web is not only one of the two best things available on the Internet (the other being the mailing lists) and what most people will use hereafter, it's also the easiest thing to use. The reason you need the separate Internet service provider, as well as the Web-browser software, is that you can't access the Web without a high-priced, high-speed link called a SLIP or PPP connection. Because those are so expensive and complicated, they're out of range for most individuals or families. However, you can get access to that kind of connection through one of the service providers. Part of what you pay them a monthly fee for is access to their SLIP or PPP connection. Not every provider has this, so be sure to ask.

# The software you need for the Net

Now for the software. Assuming you want to avoid the commercial services' Internet gateways and tolls, and get onto the Net and Web directly, there are two basic kinds, and either one will do fine.

## Specialized Net software

The first kind of program is proprietary, designed specifically for one service and usable only with that service, just as America Online, eWorld, Interchange,

Prodigy, and Women's Wire require you to use their software. Two good options here, both of which work for both PCs and Macs, are PSI Net's Pipeline and Netcom's Netcruiser, as shown in Figs. B-1 and B-2. The one drawback is that you can use Netcruiser or Pipeline with only those services, so if you decide to switch services you'll need new software.

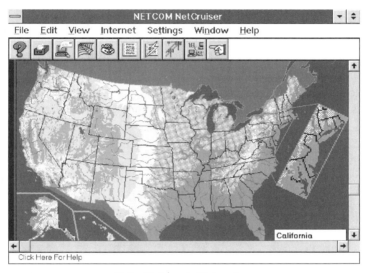

**B-1** *Netcom's Netcruiser.*

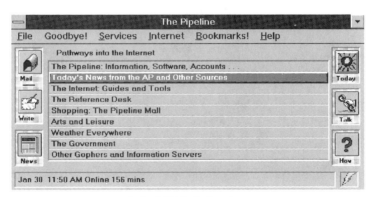

**B-2** *PSI Net's Pipeline.*

# All-purpose Net software

The second kind of software also gives you an automated, graphical interface, but works with any service. (Even though the software works with different Internet service providers, that provider still needs to give you the SLIP or PPP connection if you want to explore the Web.) This kind of software gives you more options, which also makes things more confusing. Mosaic is a free-of-charge graphical Web browser, thus that's one reasonable option, but it was the first of its kind so people have already improved on it or created similar software that's even better. Netscape is the leading Web program as this book goes to press, and has replaced the original version of Mosaic as the most popular choice.

To cut to the chase, I'd recommend that you get one of these three: Netscape, which is available for Windows and will be soon for Macs; Internet in a Box if you use Windows, which includes Air Mosaic for the Web; or OS/2 Warp if you use OS/2 rather than Windows or Mac. Netscape and Internet in a Box are shown in Figs. B-3 and B-4. All three of these programs also include software for gopher, FTP, and telnet. They aren't the only choices nor the only good programs, but they're the most popular.

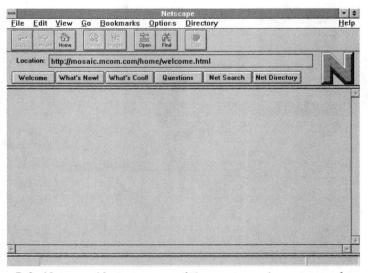

**B-3** *Netscape Navigator, one of the most popular programs for browsing the World Wide Web.*

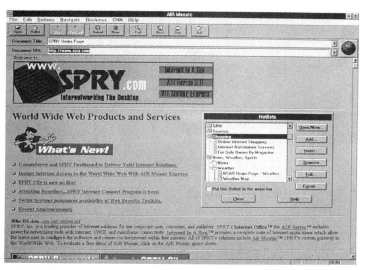

**B-4** *Internet in a Box, which includes Spry's Air Mosaic and other Internet tools.*

If you're a purist and refuse to use GUI software, or can't for some reason, you can still get on the Internet and the Web with a text-only program such as Lynx. But you won't be able to get the graphics, photographs, and sound on the Web, which is at least half of the point of being there. The real reason that the Web is so popular, though, is because it has hypertext links, which means that all you have to do is click on a word or page that's highlighted in color and it instantly takes you to that source. Well, not always instantly, but it makes the trip so effortless that it's a lot like being teleported. You can go from Peoria, Illinois to Pisa, Italy about as fast as you can say, "Beam me up, Scotty."

Please bear in mind that I'm giving you the simplified version of all this, lest your mind go numb. (I hear you: "She thinks this sounds simple? Right.") I'm also recommending more than the bare-bones options in the process, because that's the direction things are headed and where most people will ultimately want to be. So it's better to go straight to the gate than to get there by some meandering path and have to constantly buy and learn new things along the way.

The point is that whichever you choose, you need the fancier software plus the SLIP or PPP connection to use the Web, and if you're going to use the Internet you might as well use the Web. Fancy or good doesn't have to be expensive. Most major commercial services will give you their Web-browser software free, but you end up paying more because they're acting as intermediaries. Other software ranges from less than $50 to $150, but might cost you less in the long run. Also, be sure to find out the system requirements for processing speed (CPU), working memory (RAM), and hard disk storage (in megabytes) before you proceed, lest you end up with fancy software you can't run.

## Cost and other considerations

While I'm on the subject of money, typical charges for an Internet service provider are about $20 a month for up to 20 hours, or $30-plus a month for anywhere from 40 to 100 hours. But monthly rates and SLIP or PPP connections aren't the only considerations in choosing an Internet service. See the following for more that are equally important:

### Internet service provider checklist

☑ Pick a provider that offers a local phone number for access, so you don't have to pay long-distance charges on top of the access fees. If you live in a small town or rural or truly remote area, that might not be possible. Yet. But many of the services now have deals with long-distance carriers that give you access to an 800 number for a flat fee, sometimes for less than $10 a month.

☑ Choose one that has been in business for at least a year and that people in your home town recommend, whether it's a local service or a national one that offers local access. There are many incompetent people out there who have declared themselves instant experts in the hope of making a fast buck off the hype.

☑ Ask the provider to send you information on their service, and remember that first impressions are often correct.

☑ Find out how many customers they have and what's the highest-speed modem access they offer. If it's 28.8 Kbps, that's a good sign.

☑ Make sure they offer SLIP or PPP access, and make sure you know which it is.

☑ Call their tech support line after dinner one night to see how tough it is to get through and how helpful they are.

☑ Make sure they have their own server, rather than using someone else's, because your access will be faster. Just ask, "Whose server do you use?" and see what they say. If their tech staff is only marginally competent, however, that server could have a lot of downtime. So, again, rely on local recommendations from satisfied customers.

You can avoid all of this because America Online, CompuServe, Delphi, Prodigy, and others either already offer or will soon offer Web access. But with the commercial nets, you usually have to make a trade-off. Unless or until competition forces change—which it will—you either won't be able to get true full access to all services on the Net through a commercial network or you'll pay significantly more for it than you would with a separate service.

– 365 –

The Microsoft Network has vowed to change that and make everything available for one flat fee, lower than anyone else's, but as we went to press they weren't operating yet and weren't giving any details. Even if they do accomplish what they've promised, it's not likely to happen until well into 1996 and will instigate yet another fare war, so to speak, among the services. And it's anyone's guess as to how that will turn out.

So if you want to be on the Internet, not just a visitor from another service, consider the following:

➤ Try the Internet through the commercial services first. In fact, try it through each of them and see which works best for your needs and interests. All offer free trial periods.

➤ If you think you'll use the Internet often, sign up for a service that's specifically for the Internet, known generically as an *Internet service provider*.

➤ Choose an Internet service provider that can give you both Web access and a local phone number. There are hundreds, and the biggest national services also offer local numbers in many cities, with new ones added almost daily.

➤ Get automated, graphical interface software, whether it's a program for a particular service, such as PSI Net's Pipeline program or Netcom's Netcruiser, or one that will work with other services too, such as Netscape or Internet in a Box. Some of these are very easy to get started with, and practically set themselves up. They install just like any other software program. Others are trickier, sometimes much trickier.

➤ Finally, go with whatever is easiest and most affordable, all around.

# C
# Resource guide

## Online services and service providers

America Online
8619 Westwood Center Drive
Vienna, VA 22182-2285
Phone: 800-827-6364 or 703-883-1625
Information via e-mail: alt@aol.com

CompuServe
5000 Arlington Center Circle
Columbus, OH 43220
Phone: 800-848-8199 or 614-529-1212 (U.S. and Canada)
Information via e-mail: 70006.101@compuserve.com
Direct access in Argentina, Australia, Austria, Chile, France, Germany, Hong Kong,
Hungary, Israel, Japan, Mexico, New Zealand, Scandinavia, South Africa, and the U.K.

Delphi Internet Services
1030 Massachusetts Avenue
Cambridge, MA 02138
Phone: 800-695-4005 or 617-491-3342
Information via e-mail: askdelphi@delphi.com

Echo
97 Perry Street
New York, NY 10014
Phone: 212-255-3839
Modem access: 212-989-8411 or 212-898-3382 (high-speed)
Information via e-mail: info@echonyc.com

eWorld
A Computer, Inc.
P.O. Box 4493
Bridgeton, MO 63044-9718
Phone: 800-775-4556 (U.S. and Canada) or 408-974-1236
Information via e-mail: info@eworld.com

GEnie
P.O. Box 6403
Rockville, MD 20849-6403
Phone: 800-638-9636
Direct access in Canada, Japan, Germany, Switzerland, Austria, Hong Kong, and via public
dial networks in 12 other countries.

Interchange Online Network
25 First Street
Cambridge, MA 02141
Phone: 800-595-8555 or 617-252-5000
Information via e-mail: info@zdi.ziff.com

Lexis-Nexis Research Services
9443 Springboro Pike
P.O. Box 933
Dayton, OH 45401
Phone: 800-227-4908 or 513-865-6800, ext. 5858

Microsoft Network
Microsoft Corporation
One Microsoft Way
Redmond, WA 98052-6399
Phone: 206-882-8080

Netcom/Netcruiser Software
3031 Tisch Way
San Jose, CA 95128
Phone: 800-488-2558 or 408-253-3376
Information via e-mail: info@netcom.com

The Pipeline Network
150 Broadway, Suite 610
New York, NY 10038
Phone: 800-453-PIPE or 212-267-3636
Information via e-mail: info@pipeline.com

Prodigy
445 Hamilton Avenue
White Plains, NY 10601
Phone: 800-Prodigy (U.S. and Canada)
Information via e-mail: info@prodigy.com

The WELL
27 Gate Five Road
Sausalito, CA 94965
Phone: 415-332-4335
Modem access: 415-332-6106
Information via e-mail: info@well.com

Women's Wire
1820 Gateway Boulevard-150
San Mateo, CA 94404
Phone: 800-210-8998 or 415-378-6500
Information via e-mail: info@wwire.net

# Software

HiJaack Graphics Suite
Inset
71 Commerce Drive
Brookfield, CT 06804-3405
Phone: 800-374-6738 or 203-740-2400

Internet in a Box
Spry
316 Occidental Avenue South
Seattle, WA 98104
Phone: 800-557-9614, ext. 214 or 206-447-0300
Information via e-mail: iboxinfo214@spry.com

Journalist
PED Software Corporation
10101 N. DeAnza Boulevard, #400
Cupertino, CA 95014
Phone: 408-253-0894

Norton Antivirus
Symantec Corporation
10201 Torre Avenue
Cupertino, CA 95014-2132
Phone: 800-441-7234 or 408-253-9600

WinComm Pro & WinFax Pro
Delrina
895 Don Mills Road, 500-2 Park Centre
Toronto, ON M3C 1W3
Canada
Phone: 416-441-3676 or 408-363-2345

# Hardware

NEC Technologies
1255 Michael Drive
Wood Dale, IL 60191
Phone: 800-NEC-INFO (U.S. and Canada)
MultiSync monitors and NEC MultiSpin CD-ROM drives

Creative Labs
1901 McCarthy Blvd.
Milpitas, CA 95035
Phone: 408-428-2329
Sound Blaster 16 system

Apple Computer, Inc.
One Infinite Loop
Cupertino, CA 95014
Phone: 415-996-1010

Practical Peripherals
375 Conejo Ridge
Thousand Oaks, CA 91361
Phone: 805-497-4774
Modems

Toshiba America Information Systems
Computer Systems Division
9740 Irvine Boulevard, P.O. Box 19724
Irvine, CA 92713
Phone: 800-334-3445 or 714-583-3000

U.S. Robotics
8100 N. McCormick Boulevard
Skokie, IL 60076-2920
Phone: 708-676-7113
Modems

Western Digital
8105 Irvine Center Drive
Irvine, CA 92718
Phone: 800-568-9272 or 714-932-5000
Hard drives

# D
# Comparison of online services

| | America Online | CompuServe | Delphi | GEnie | Prodigy | Internet |
|---|---|---|---|---|---|---|
| **Owned by** | Public stock | H&R Block | News Corp. | GE | IBM-Sears | Various & taxpayers |
| **Number of members** | 2.5 million | 3 million | 100K+ | 400,000 | 2 million | 6–20 million |
| **Percent women** | 30% | 12–17% | 20% | 23% | 40% | 15–30% |
| **International** | Link in 1995 | 148 countries | UK | No | No | 200 countries |
| **Forums free** | Most | Help only | All | Most | Most | Yes |
| **Free e-mail** | Yes | 90 messages | Yes | Yes | No | Yes |
| **Charge Net mail** | No | Yes, incoming | No | No | Yes | No |
| **For women** | 1 section, Forum '95 | 1 forum, 4 sections | 1 forum | 2 sections | 1 forum + $ | 3 newsgroups+ |
| **Worldwide Web** | Yes | Yes | Due 1995 | Planned | Yes | With software & access |
| **Internet access** | Partial | Full, Web '95 | Full, Web '95 | 1995 TBA | Partial w/Web | It *is* the Internet |
| **Free trial period** | 10 hours | 1 month, part | 5 hours | 4 hours | 1 month | N/A |
| **Hours free/month** | 5 | | 4 or 20 | 4 | 5 | 20 = typical |
| **Day rates higher** | No | No | Yes | Yes | No | No |
| **Basic rate/mo.*** | $9.95 | $9.95 | $10 or $20 | $8.95 | $9.95 | $17.50 average |
| **Rate/hour @ 14.4** | $2.95 | $4.80 | $4 or $1.20 | $2.00 | $2.95 | varies |
| **Highest baud** | some 28.8 | 28.8 | 14.4 | 9600 | 14.4 | 28.8+ |
| **Minimum baud** | 2400 | 300 | 1200 | 2400 | 2400 | 2400 |
| **Automated software** | Windows/Mac | Win, Mac, OS/2, DOS | Windows | Win, Mac | Win, Mac | Available |
| **Offline readers** | None | Win, Mac, DOS | Win, DOS | DOS | None | Partially available |
| **Software cost** | Free | Free | Free | Free | Free | Free, up to $100 |
| **Threaded messages** | No | Yes | Yes | No | No | No |
| **Send faxes** | Yes | Yes | Yes | Yes | Yes | No |
| **PCs or Macs** | Windows/Mac | Yes | Yes | Yes | Yes | Yes |
| **Amiga, Tandy, etc.** | No | Manually | Manually | Manually | No | Manually |
| **Ads displayed** | Some/notices | No | No | No | Yes | On Web |
| **Research databases** | Limited | Excellent | Limited | Good | Limited | Excellent |

| | ECHO | eWorld | Interchange | MS Net work | The WELL | Women's Wire |
|---|---|---|---|---|---|---|
| Owned by | Private co. | Apple Computer | AT&T | Microsoft | Private co. | Private company |
| Number of members | 3,500 | 100,000 | New 1995 | New 1995 | 11,000 | 1,500 |
| Percent women | 38% | 20% | New 1995 | New 1995 | 20% | 90% |
| International | By Telnet | 60 countries | Europe Online | 43 countries | By Telnet | No, planned |
| Forums free | Yes | Yes | TBA | No | Yes | Yes |
| Free e-mail | Yes | Yes | Yes | Yes | Yes | Yes |
| Charge Net mail | No | No | TBA | No | No | No |
| For women | 3 forums | 1 forum | No | 2 forums | 2 forums | All |
| Worldwide Web | Available | Due 1995 | Planned | Late 1995 | Late 1995 | No |
| Internet access | Full | Due 1995 | Due 1995 | Late 1995 | Full | Selected services |
| Free trial period | 1 month | 10 hours | Yes, TBA | Yes, TBA | No | 10 hours |
| Hours free/month | 30 | 4 in U.S., 1 abroad | TBA | TBA | 0 | 0 |
| Day rates higher | No | No | No | No | No | No |
| Basic rate/mo.* | $19.95 | $8.95 | TBA | $10 | $15 | $9.95 |
| Rate/hour @ 14.4 | N/A | $2.95 | TBA | TBA | $2 @ 9600 | $3.95 |
| Highest baud | 28.8 | 14.4 | 14.4 | 14.4 | 14.4 | 14.4 |
| Minimum baud | N/A | 2400 | 9600 | 2400 | 2400 | 2400 |
| Automated software | No | Yes | Yes | Yes | Late 1995 | Yes |
| Offline readers | No | Mac, Win due | Win | Win95 | TBA | Win/Mac |
| Software cost | N/A | Free | Free | MSWin95 | TBA | Free |
| Threaded messages | Yes | No | Yes | Yes | Yes | No |
| Send faxes | No | Yes | Yes | No | No | No |
| PCs or Macs | Both | Mac only | Windows only | Win95 only | Manually | Both |
| Amiga, Tandy, etc. | Manually | No | No | No | Manually | No |
| Ads displayed | No | No | TBA | TBA | No | No |
| Research databases | None | None | TBA | TBA | None | None |

* Usually slightly higher outside continental U.S.

# E

# Nonprofit organizations online

## On the World Wide Web

The following resources are from the directory Nonprofit Organizations on the Internet, 1995, by Ellen Spertus. Also see The Internet Nonprofit Center at http://human.com/inc.

Al-Anon and Alateen
http://solar.rtd.utk.edu/~al-anon/

The Alliance for Public Technology
http://apt.org/apt.html

Alternet
http://www.igc.apc.org/an/alternet.html

America Responds to AIDS
http://bianca.com/lolla/politics/aids/aids.html

American Academy of Pediatrics
http://www.aap.org/dogl/dogl.html

American Civil Liberties Union
gopher://pipeline.com/11/society/aclu

The American Indian College Fund
http://hanksville.phast.umass.edu/defs/independent/AICF.html

The American Red Cross
http://www.crossnet.org/

Amnesty International Online
http://www.io.org/amnesty/overview.html

ASHA
http://theory.stanford.edu/people/katiyar/asha.html

AEGEE
http://www.uni-konstanz.de/studis/aegee/index.html

The Association of America's Public Television Stations (APTS)
http://www.universe.digex.net/~apts

Australian Consumers' Association
http://www.sofcom.com.au/ACA/

Believe in Yourself
http://drum.ncsc.org/~carter/selfdex.html

The Benton Foundation
http://www.cdinet.com/Benton/home.html

Bethany Christian Services
http://www.bethany.org/bethany/what_we_do.html

Board of European Students of Technology (BEST)
http://www.nada.kth.se:80/~ovidiu/best/

Canine Companions for Independence
http://grunt.berkeley.edu/cci.html

Cannabis Action Network
http://bianca.com/lolla/politics/can/can.html

Center for Civic Networking
http://www.civic.net:2401/

Center for Civil Society International
http://solar.rtd.utk.edu/~ccsi/brochure.html

The Center to Prevent Handgun Violence
http://bianca.com/lolla/politics/handguns/handgun.html

CREST, the Center for Renewable Energy & Sustainable Technology
http://solstice.crest.org/common/crestinfo.html

Center for World Indigenous Studies
http://www.halcyon.com/FWDP/fwdp.html

Child Quest International
http://www.omega.com/adima/bands/child_quest/cqmain.html

Child Relief & You (CRY)
http://www.acsu.buffalo.edu/~kripa/cry.html

Circle K International
http://nyx10.cs.du.edu:8001/~jwolff/cki.html

Cocaine Anonymous
http://www.ca.org

Communications for a Sustainable Future
gopher://csf.colorado.edu/

Computer Professionals for Social Responsibility
http://www.cpsr.org/home

The Concord Coalition
http://sunsite.unc.edu/concord/

David LaMacchia Defense Fund
http://www-swiss.ai.mit.edu/dldf/home.html

Electronic Frontier Foundation
http://www.eff.org/

EnviroWeb
http://envirolink.org:/start_web.html

The Environmental Hazards Management Institute
http://www.ehmi.org/

Families Against Mandatory Minimums
http://bianca.com/lolla/politics/famm/famm.html

Feminists for Life of America
http://www.cs.indiana.edu/hyplan/ljray/lifelink/plfem.html

F.A.C.T.Net, Inc. (Fight Against Coercive Tactics Network)
http://www.acmeweb.com/factnet/

EXTRA! the magazine of FAIR (Fairness & Accuracy In Reporting)
http://theory.lcs.mit.edu/~mernst/fair

The Foundation Center
http://fdncenter.org/

The Foundation for National Progress
http://www.mojones.com/mojo_info.html

Free Burma
http://www.interactivist.virtualvegas.com/freebrma/freebrma.htm

The Freedom Forum
http://www.nando.net/prof/freedom/1994/freedom.html

Friends of the Earth
http://www.foe.co.uk/

George Lucas Educational Foundation, Edutopia project
http://glef.org

Global Fund for Women
http://www.ai.mit.edu/people/ellens/gfw.html

Greenpeace
http://www.greenpeace.org/

Gurukul
http://www.acsu.buffalo.edu/~naras-r/gurukul.html

Hindu Heritage Endowment
http://www-bprc.mps.ohio-state.edu/cgi-bin/hpp?hhe1.html

Home Recording Rights Coalition
http://www.digex.net/hrrc/hrrc.html

House Rabbit Society
http://www.psg.lcs.mit.edu/~carl/paige/HRS-home.html

Human Services Research Institute
gopher://ftp.std.com/11/nonprofits/hsri

HungerWeb
http://www.hunger.brown.edu/hungerweb/

The Immunization Action Coalition
http://www.winternet.com/~immunize

INFACT
http://sunsite.unc.edu/boutell/infact/infact.html

Institute for Global Communications (IGC) Networks
http://www.igc.apc.org/index.html
This contains PeaceNet, EcoNet, ConflictNet, LaborNet, etc. The site has an excellent,
extensive list of women's organizations from all over the world who are online through
IGC, with e-mail addresses.

The International Institute for Sustainable Development
http://www.iisd.ca/linkages/index.html

The International Marinelife Alliance (IMA)
http://www.actwin.com/fish/ima

The International Service Agencies
http://www.charity.org/

Join Together
gopher://gopher.jointogether.org:7003

Lead... or Leave
http://www.cs.caltech.edu/~adam/lead.html

League of Conservation Voters
http://www.econet.apc.org/lcv/lcv_info.html

The League for Programming Freedom
http://www.lpf.org/

The Marine Mammal Center (TMMC)
http://www.spiderweb.com/sw/mmc.html

The Milarepa Fund
http://bianca.com/lolla/politics/milarepa/milarepa.html

National Center for Missing and Exploited Children
http://www.scubed.com:8001/public_service/missing.html

The National Child Rights Alliance
http://NCRA/ncra.html

National Rifle Association
http://www.nra.org

N¦USA
http://www.charities.org/

The Anthony Nolan Bone Marrow Trust
http://wombat.doc.ic.ac.uk/bone-marrow.html

North American Institute
http://sol.uvic.ca:70/0h/nami/HOMEPAGE.html

Peace Corps Web
http://www.intac.com/PubService/peace-corps/peace-corps.html

Plugged In
http://server1.imi.com/ims/pi/PluggedIn/PluggedIn.html

Precious in HIS Sight Internet Adoption Photolisting
http://www.gems.com/adoption/

Prison Legal News
http://PLN/pln.html

The Privacy Rights Clearinghouse
http://www.manymedia.com/prc/index.html

Project America
http://www.mit.edu:8001/activities/project-america/homepage.html

Project Vote Smart
gopher://chaos.dac.neu.edu/11/pvs-data

Queer Resources Directory
http://vector.casti.com/QRD/.html/QRD-home-page.html

ReliefNet
http://www.earthweb.com:2800/reliefnet.html

Rock for Choice
http://bianca.com/lolla/politics/rockforchoice/r4c.html

Rotary Foundation
http://www.tecc.co.uk/public/PaulHarris/foundatn.html

The Seva Foundation
http://www.well.com/Community/Seva/

SOS China Education Fund
http://ifcss.org:8001/www/pub/org/sos-edu/.index.html

Sierra Club
http://www.sierraclub.org/

Stop Prisoner Rape
http://www.ai.mit.edu/people/ellens/SPR/spr.html

Street Project
http://www.phantom.com/~gerster/home.html

Student Environmental Action Coalition
http://bianca.com/lolla/politics/seac/seac.html

The Vera Institute of Justice
http://broadway.vera.org/

The World Affairs Council of Philadelphia
http://libertynet.org/business/econ-dev/ucsc/wac/wac.html

World Conservation Monitoring Center
http://www.wcmc.org.uk/

The Coalition to Ban Dihydrogen Monoxide
http://www.circus.com/~no_dhmo/

People for the Ethical Treatment of Software
http://paul.spu.edu/~zylstra/comedy/computer/watch-list.txt

# Members of the Organization Access Conference on Women's Wire

ACES, Children for Enforcement of Support

Alumnae Resources, Career Development and Planning

Boston Women's Health Book Collective (*Our Bodies, Ourselves*)

CAPP, Washington Feminist Faxnet

Career Action Center, Career Center in the South Bay (San Francisco)

Commission for Sex Equity

CMCR, Mother & Child Rights

Democratic Activists for Women Now (DAWN)

Domestic Abuse Awareness Project (DAAP)

Equal Rights Advocates (ERA)

Fund for the Feminist Majority

Gift From Within, Survivors of Post-Traumatic Stress

Media Alliance

National Education Center for Women in Business (NECWB)

National Organization for Women (NOW)

Planned Parenthood Federation of America

San Francisco Commission on the Status of Women

Speak-Up, Leadership Education for Girls

Students Organizing Students (SOS)

Women in the Fire Service

The Women's Building

Women's Cancer Resource Center

# Organizations with information on Prodigy's Women's Leadership Conference

American Medical Women's Association

AWED (American Woman's Economic Development Corporation)

Center for Creative Leadership

Center for the American Woman and Politics

Center for Policy Alternatives

Commonwealth Fund Commission on Women's Health

Eric Digests

Federal Agencies

    Bureau of the Census

    Centers for Disease Control and Prevention

    Consumer Product Safety Commission

    Department of Education

    Department of Health and Human Services

    Department of Justice, Bureau of Justice Statistics

    Department of Labor Women's Bureau

    Equal Employment Opportunity Commission

    National Archives

    National Center for Education Statistics

    National Center for Health Statistics

    National Institutes of Health

    Office of Personnel Management

    Small Business Administration, Office of Women's Business Ownership USIA

Female members of the U.S. Congress, House and Senate

Girl Scouts, U.S.A.

Institute for Women's Policy Research

International Women's Health Coalition

Leadership America
National Foundation for Women Business Owners
National Museum for Women in the Arts
National Parent Information Network
National Women's Business Council
National Women's Health Resource Center
National Women's History Project
National Women's Political Caucus
Religious Coalition for Reproductive Choice
The White House
The Women's Center
U.N. Fourth World Conference on Women
Winning Strategies Magazine
Women Today Magazine
Women Work
Women's Network of NCSL (National Conference of State Legislatures)
Women's Publications, Inc.
World Health Organization Global Commission on Women's Health

Many business or professional organizations also have sections within relevant forums or online liaisons. They're more often found on the commercial networks, primarily CompuServe.

# F
# Media online

## Magazines

Citations, abstracts, or full-text articles from many of these magazines are also in ZiffNet's Magazine Database Plus on CompuServe and Prodigy, in the Magazine Index within Knowledge Index on CompuServe, in UnCover on the Web, or through the News database on Nexis. Those listed here sponsor their own forums on the networks indicated. New publications are added monthly.

*Ad Age*, Prodigy

*American Woodworker*, America Online

*Atlantic Monthly*, America Online

*Backpacker*, America Online

*Bicycling*, America Online

*Boating*, America Online

*Bon Appetit*, Epicurious on the Web

*Business Week*, America Online

*Car and Driver*, America Online

*Christianity*, America Online

*Christian Reader*, America Online

*Crain's Small Business*, America Online

*Consumer Reports*, America Online, CompuServe, Prodigy

*Disney Adventures*, America Online

*Elle*, America Online

*Entrepreneur*, Entrepreneurs Forum on CompuServe

*Family Handyman*, CompuServe

*Fortune*, CompuServe

*Flying*, America Online

*Home*, America Online

*Home Office Computing*, America Online and selected articles on Prodigy

*InfoWorld*, eWorld, with CompuServe to come

*Inside Media*, America Online

*Investor's Business Daily*, America Online

*Kiplinger's*, Prodigy

*Longevity*, America Online

*MacWorld*, America Online and eWorld

*Mobile Office*, America Online

*Mother Jones*, Mother Jones Interactive on the Web

*Ms.*, Echo

*National Geographic*, America Online, plus Activities section of Just for Kids on Prodigy

*The Nature Conservancy*, America Online

*The New Republic*, America Online

*The New York Times*, TimesFax on the Web

*New Woman*, the Relationships section on Women's Wire

*Newsweek*, Newsweek Interactive on Prodigy

*Omni*, America Online

*PC/Computing*, CompuServe

*PC Magazine*, Interchange

*PC Novice*, America Online

*PC Today*, America Online

*PC World*, America Online and CompuServe

*Playbill*, Playbill Online on Prodigy

*People*, CompuServe

*Popular Photography*, America Online

*Road & Track*, America Online

*Rolling Stone,* CompuServe

*Self*, on the Web

*Smithsonian*, selected information in AOL's Electronic University

*Spin*, America Online

*Stereo Review*, America Online

*Time*, America Online

*Today's Christian Woman*, America Online

*Travel & Leisure*, America Online

*Travel Holiday*, America Online

*Saturday Review* (archives), America Online

*Scientific American*, America Online

*Sports Illustrated*, CompuServe

*Sports Illustrated for Kids*, Prodigy

*US News & World Report*, CompuServe

*Wired*, America Online (Hotwired on the Web)

*Writer's Digest*, eWorld

*Woman's Day*, America Online

*Worth*, America Online

– 389 –

# Collections from magazine publishing groups

The following are due sometime in 1995: Condé Nast's Epicurious site on the Web, which is likely to include articles from *Bon Appetit* and *Gourmet*; and Hearst's Home Arts, also on the Web and possibly one of the commercial networks, with articles from *Good Housekeeping*, *Colonial Homes*, *Victoria*, *Popular Mechanics*, *Smart Money*, and *House Beautiful*, plus other information and events. Also see Hearst's Multimedia Newstand on the Web.

# Major Newspapers
## (some bysubscription only after trial)

*Atlanta Journal-Constitution*'s Access Atlanta on Prodigy

*Chicago Tribune*'s Chicago Online on America Online

*Detroit Free Press*, CompuServe

*The Los Angeles Times*' TimesLink, Prodigy

*New York Times*' @Times (selected same-day stories only) on America Online and TimesFax on the Web

*Newsday* Direct on Prodigy

*San Jose Mercury News*' Mercury Center on America Online

*The Washington Post*'s Digital Ink on Interchange

# TV stations and networks

ABC's News on Demand on America Online

CNBC on Prodigy

CNN, the Cable News Network, on America Online and CompuServe

The Discovery Channel on America Online

Nickelodeon on Prodigy

NOVA and WGBH/Boston on Prodigy's Just for Kids Activities section

# G

# How to find people and places online

## On the Internet

There are innumerable lists or directories and many *search engines*, as they're called, which allow you to enter keywords to have the search engine check the entire Internet—theoretically—for what you want, and report back lickety-split. The ones listed here, however, are among the best. No one source has everything. It's just too vast and complex. So it pays to check at least a couple of sources, because you'll invariably turn up a few things on one that you didn't find on another.

## General or comprehensive sources

**The Clearinghouse for Subject-Oriented Resource Guides**
The University of Michigan's Library and School of Information and Library Science covers gophers, telnet, and FTP sites, newsgroups, mailing lists, phone numbers, and more. You'll find excellent women's studies and women's health information and leads here.
Gopher: gopher.lib.umich.edu.
FTP: una.hh.llib.umich.edu.
Telnet: una.hh.lib.umich.edu 70.
Web: http://www.lib.umich.edu/chhome.html or http://una.hh.lib.umich.edu/11/inetdirs.

### Infoseek
An affordably priced, fee-based search tool for the Web and Internet newsgroups, at http://www.infoseek.com.

### Lycos
Another Web search engine (named after a spider) that searches for keywords.
Web: http://lycos2.cs.cmu.edu/cgi-bin/pursuit

### Starting Points for Internet Exploration
Web: http://www.ncsa.uiuc.edu/SDG/Software.Mosaic/SP
/NetworkStartingPoints.

### Yahoo: A Guide to the World Wide Web
Another search engine, with more than 15,000 links, including some interesting ones you're not likely to find elsewhere. Especially good for the arts and rather off-center sites. Searchable by keywords, with links to other search engines. Great resource but no longer free after the trial period, alas (see Fig. G-1).
Web: http://akebono.stanford.edu/yahoo

*Yahoo*

[ What's New? | What's Cool? | What's Popular? | A Random Link ]

| Y Top | Up | Search | Mail | Add | Help |

- Art(586) NEW
- Business(8211) NEW
- Computers(3208) NEW
- Economy(871) NEW
- Education(1784) NEW
- Entertainment(8480) NEW
- Environment and Nature(262) NEW
- Events(63) NEW
- Government(1209) NEW
- Health(528) NEW

**G-1** *Yahoo, one of the best ways to find things on the Web.*

### The Yanoff List of Special Internet Connections
Another popular search tool. Indexed by topic, from agriculture to weather.
Web: http://www.uwm.edu.Mirror/inet.services.html

# Gopher sites

### Gopher Jewels
A gopher best-of list.
Web: gopher://cwis.usc.edu/11/Other_Gophers_and_Information_Resources
—> Gophers_by_Subject —> Gopher_Jewels

### The RiceInfo Gopher
Rice University, Houston, Texas
Indexed by subject, from aerospace to weather.
Gopher: riceinfo.rice.edu
Web: gopher://riceinfo.rice.edu:70/11/subjectnetworks

# Mailing lists

### The List of Lists from SRI (Stanford Research Institute)
The best of the two best ways to find out what mailing lists exist. Caveat: uncompressed, it takes up about 1.5 megs on your hard drive. This means it's long, really long. Send e-mail to mail-server@sri.com, with the subject line blank and body of message as SEND interest-groups. Or you can FTP to sri.com/netinfo/ —> interest-groups.

### Publicly Accessible Mailing Lists
Also known as the PAML list, this one covers only Usenet lists, not Bitnet, but just sorting through it will keep you busy in your spare time for weeks, maybe months. Fortunately, it's broken up into several files, so you can try one and see if it's to your liking. Send e-mail to mail-server@rtfm.mit.edu, with the body of message as SEND usenet/news.answers/mail/mailing-lists/. FTP: rtfm.mit.edu/pub/usenet/news.answers/mail/mailing-lists/

# Newsgroups

### Lists of newsgroups
Web: http://www.cis.ohio-state.edu:80/hypertext/faq/usenet/active-newsgroups/top-html
FTP: ftp.uu.net/usenet/news.answers/active-newsgroups

### Newsgroups about newsgroups
Try these: news.lists, news.groups, news.announce.newgroups, or news.answers.

**Newsgroup search services**

If you don't have time to keep up with the newsgroups but don't want to miss discussions of a particular topic, there are a couple of free services that will search all newsgroup content for you and alert you to anything that comes up according to your instructions. This process doesn't work very well, but perhaps it could with patience and fine-tuning. The first is Oracle, which—as in the ancient days of Greece—will give you an answer to any question you ask. This isn't because it's omniscient, but because it has a sophisticated program that searches all Usenet newsgroups for mention of the terms in your request. Sometimes, though, you'll get a humorous response. To find out more, send e-mail to oracle@cs.indiana.edu with the subject line HELP. The second, Stanford Netnews' Filtering Service, searches Usenet newsgroups on the Internet for your choice of keywords, then sends anything relevant it finds to you by e-mail. Another great idea that, so far, returns more extraneous than relevant information. Try tinkering with the few commands or getting tech support.

# World Wide Web

**EINet (Enterprise Integration Technologies Web Resources)**
Web: http://www.eit.com/web/web.html

**Inter-Links**
Another index, and a somewhat quirky one at that. Its author calls it "ego-ware," but in a likable manner.
Web: http://alpha.acast.nova.edu.start.html

**The World Wide Web Worm**
One of the main search engines on the Web.
Web: http://www.cs.colorado.edu/home/mcbryan/WWWW.html

# Search methods on the Internet

The three basic ways to find people on the Internet are as follows:

***By the person's name only*** Send e-mail to netaddress@nri.reston.va.us. Leave the subject blank and put the person's name in the body of the message. (This sometimes works and sometimes doesn't, which is the way of things on the Net.)

***By the person's name and their school or organization*** Try Netfind User Lookup. Telnet to telnet://bruno.cs.colorado.edu.

**With only part of the address** WhoIs helps you find people, sometimes, if you know at least part of their e-mail address, such as mit.edu, which stands for MIT, or the Massachusetts Institute of Technology. To try it, type whois mail service@rs.internic.net and use the subject line HELP.

# On commercial services
## America Online

To find places on America Online, click on Go To on the menu bar at the top of the screen, and choose (click to highlight or activate) whether you want to search by service name, keyword, search words, menu path, or service description. Then just type in a word, topic, or name, and choose where you want to go from the menu that shows up after that. Many times a logical keyword gets you nowhere on AOL, alas, even though there's something on the service related to that term. *Sex* won't turn up anything, for instance, although *sexuality* will take you to the health forum. To find people, check the directory under the Members menu item.

– 395 –

## CompuServe

Click on Find or the question mark on the menu bar at the top of the screen in CompuServe Information Manager, or type find in text-only mode, followed by a keyword or phrase. It will usually show you what's where if you come anywhere close to a related word. There's also a feature called File Finder that searches the entire system for a particular file if you know the filename or keywords included in the description, or in the case of PC files, the file extension. When I was looking for photos for this book, for instance, I used TIF, EPS, and GIF—all file formats for photos or graphics—to find everything in a forum's library or throughout the whole network. To find people, use the last item under the Mail menu, which is the Member Directory. When you click on it, a box pops up for you to enter the person's name, and the computer does the rest. If you're online manually, type FIND members.

# Delphi

Delphi was creating new software when this book went to press, so look for the Help instructions or call tech support.

# eWorld

Go to the Information Center building, then click on Find at the left of the screen and type in a keyword or phrase. To find people, check the Post Office, as though you were going to send someone e-mail.

# GEnie

Click on the Pages icon and type in the name of the Roundtable or section or a keyword. GEnie is also currently changing so these instructions will change too. To find people, look under the Members menu.

# Prodigy

Click on the letter J (for *jump*) on the menu bar at the bottom of the screen, then type in a keyword or phrase. To find people, jump to the Member Services menu.

# Women's Wire

Look for the Map listings at the bottom of the directory screens. You can print the maps or save them to a file on disk for later reference. To find people, open a New Message under the Mail menu, click on the To line, and type in part of the name of the person. A list of matches will pop up.

# Glossary

## Terms

**agents** Similar to knowbots or bots.

**bandwidth** The capacity of any online system to transmit data. It's finite, thus clogging it with long monologues or meaningless messages or those too short to warrant a message at all, such as "I agree" or "Thanks," is commonly called "a waste of bandwidth."

**BBS** Electronic bulletin board system. Usually refers to local- or regional-only networks, but forums on major networks are also electronic bulletin boards, technically.

**bot** Short for knowbot.

**bozo filter** A way to screen out the jerks. Also known as a squelch feature.

**clueless newbie** Somebody new on the Net who didn't bother to read the FAQ file or get oriented to the culture and netiquette before foisting themselves on the masses.

**commercial service** Online services that are in business to make a profit, such as America Online, CompuServe, and Prodigy; they have Internet addresses with .com at the end. Actually, everybody running any online network hopes to profit from it somehow, particularly those on the World Wide Web, which is part of the Internet, which is people often think is free, which is a fallacy. Somebody has to pay for all of it, whether it's through taxes, tuition, or upfront fees.

**cyberspace** The void within which modems help people communicate. It's also the same place you're communicating in when you talk on the phone, so the mystique is highly over-rated, but it's a useful term. Coined by William Gibson, author of the techno-fantasy book *Neuromancer*, who says he's not online.

**dial-up access** Refers to direct access through your own Internet service provider rather than getting to the Net through the gateway of a commercial service.

**domain name** The part of an Internet address that comes after the @ sign. The part that identifies what kind of address it is, such as .edu or .org or .com, is called the zone.

**e-mail** Electronic mail. You know this, right?

**emoticons** Smileys :-) and other cutesy typographical symbols used online to convey emotion. Corny, yes, but they can help prevent hurt feelings and misunderstandings.

**FAQ** Short for Frequently Asked Questions. Read the FAQ files on the Internet when you're new to a newsgroup and don't want to irritate people by asking questions that have already been answered a thousand times (which will prompt some people to accuse you of being a clueless newbie). It will also help you get oriented and feel more sure of yourself in posting public messages.

**flame** An angry or hostile message or attack on someone. Think *incendiary* and *flame-thrower*, and you'll get the idea. Most *flamers*, as they're known, are male, and most of them are young or immature.

**FTP** File transfer protocol. Telling your computer to go inside another computer that's linked to the Internet, find a file, and send it back to your computer.

**GIF** A certain kind of graphics file, which can be a photo or an illustration. These require special readers, but those are built into many proprietary software programs now. Also see JPEG.

**gopher** A software tool to search for things on the Internet.

**GUI** Graphical User Interface, which means software that has icons or pictures to click on, and clicking on them automates complicated commands you never have to bother learning. The alternative is a command-line interface, which means strictly text, and you have to learn the commands.

– 399 –

**hacker** Computer-savvy people able to "hack out" a solution to a problem.

**home page** The first screen you come to at a Web site.

**hypertext links** Think "fast connections to words." Clicking on the button, icon, or word that activates a hypertext link takes you to another source in another place, anywhere from instantly to after a few minutes, depending on how busy the system is and how many graphics there are to transmit. Hypertext links are one of the best things about the World Wide Web, but are now available on some commercial networks, such as the Interchange.

**Internet** The international network of networks started by the U.S. Government as a way to circumvent cut-off communications in the event of a major war. Evolved into a way for researchers to communicate, and now allows all of us to communicate, including those who would be cut off from communicating otherwise because of a coup, war, or revolution. Not synonymous with *online* because you can be online, say on Prodigy, but not on the Internet unless you go there through their gateway.

**IRC** Internet Relay Chat. The Internet's version of chat channels, an online version of ham or CB radio. Enables you to chat in real time.

**IRQ** Rather complicated settings within PC-type computers, meaning those that use DOS, Windows, or OS/2 operating systems. Can affect your modem installation and has to do with how your computer conveys commands to itself.

**JPEG** Another kind of graphics file format, similar to GIF.

**knowbot** Same as *bot*. A computer script that automates commands, telling the computer to search many different sites or sources for information. Among other uses, this is how some electronic clipping services find the news from many different publications and deliver it to you in capsule form.

**log on** or **log in** Connect to an online service via modem.

**lurk** To read messages without replying or participating in the discussion. This is a good way to get oriented, but not a good way to get the full benefit.

**mailing list** Messages distributed by group e-mail through the Internet. You have to sign up for these, but some are very good because they stay more focused than newsgroups.

**modem** The device that connects you to an online service through your phone line.

**MUDs** Multiuser dimensions, also known as multiuser dungeons, from the Dungeons & Dragons heritage. Places where people create virtual or simulated worlds or communities with made-up characters, including people and other creatures, complete with objects and terrain. Highly addictive for some people.

**Net** With a capital N, the Internet. With a lowercase letter, everything online.

**netiquette** Online etiquette.

**newsgroup** The equivalent of a forum on the Internet.

**online** Not synonymous with the Internet, because it also includes the commercial services, research databases such as Nexis or Dialog, and BBSs.

**real time** Right now. Typically, communicating with someone online while you're both connected, typing back and forth. If you pay a flat fee per month for unlimited hours and feel like spending your time this way, it's good. If you're paying by the hour or minute, it isn't smart unless you're independently wealthy.

**shareware** Try-before-you-buy software that requires a registration fee if you continue to use it after a trial period, which is usually 30 days. Not to be confused with freeware or public-domain software, which are both free. Shareware is not free, and if you use it without paying for it you're violating international copyright law. It's tempting because there's lots of shareware online, but it's a poor example to set for your kids and creates bad karma.

**SIG** Special-interest group. Same thing as a forum or a newsgroup, or a mailing list for that matter.

**SLIP** or **PPP** The kind of connection you need from your Internet access provider to be able to use all of the features on the World Wide Web. Also see TCP/IP.

**snail mail** Regular postal mail, also called earth mail.

**spam** To bombard people with junk e-mail or newsgroup messages, usually advertising, as in "He spammed the Net."

**squelch features** See bozo filter.

**TCP/IP** The kind of connection you need to be able to access the Internet directly, rather than through a commercial network. Also see SLIP or PPP.

**telnet** To go to a remote location, by modem, to use their computer system and access their files.

**TOS** Terms of service. You'll hear a lot of grousing, particularly from women, on America Online about people violating the network's policies, or terms of service. Occasionally it's legit; usually it's just from people who think everyone should play by their rules.

**UART** A computer chip inside your computer, or on your modem if you have an internal modem. The thing to remember is that you need an up-to-date 16650A UART if you're installing a high-speed modem, which means anything 9600 bps or faster.

**URL** Uniform resource locator, or the technical name for an address on the World Wide Web. It's called uniform because Web addresses all start with http://, which stands for hypertext transfer protocol.

**virtual** Simulated, as in virtual reality, or almost real. A virtual corporation is real, but the employees or partners are located in different places and they conduct much of their business by modem.

**virus** A computer program that makes your computer sick because it destroys or corrupts the data you've stored in it. Viruses are more likely to circulate on the Internet or be transferred in executable files—programs—rather than by straight text. The fact that they exist is a good reason to get and use an antivirus program daily, or install it as a TSR (terminate-and-stay-resident) program that's always active.

**Web browser** Software that allows you to use the full capabilities of the Web.

**Web site** A location or sponsored service on the World Wide Web.

**wired** Meaning you're online or connected to the Internet in some way, which is where Women's Wire gets its name. Also the name of a self-consciously hip magazine about related technology with 85 percent young male subscribers, but sometimes a good read, nonetheless.

**World Wide Web** Should be written Worldwide Web, but techies, not copy editors, make up such terms. The part of the Internet that is easiest to use because it has buttons and graphics to click on and hypertext links. It gives you access to much of the rest of the Internet, and makes what else is there much easier to find.

# Acronyms

The following are commonly used abbreviations for clichés, some of which help make messages diplomatic:

**BTW** By the way

**FWIW** For what it's worth

**IMO** or **IMHO** In my opinion or in my humble opinion

**IRL** or **RL** In real life or real life, a.k.a. offline

**ROFL** Rolling on the floor, laughing (more likely sitting, chuckling and considered contrived and corny)

**RTFM** Read the friggin' manual

**WYSIWYG** What you see is what you get, meaning don't expect magic, because what you see on the screen is probably representative of the real thing. Related to GIGO, or garbage in, garbage out, meaning computers, like humans, have their limitations.

# Bibliography

## Articles

Roundtable discussion. 1994. "The cutting edge: women and computing, nature or nurture?" e-mail roundtable discussion. *Los Angeles Times*, April 11, Business, Part D2: 16.

Balka, Ellen. 1991. "Women's access to on-line discussions about feminism." Computer Professionals for Social Responsibility Internet Library.

Broadhurst, Judith. 1994. "Hardwear: Mobile technology lets us mesh our jobs with our lives. But do our jobs now dominate, getting in the way of personal and family relationships?" *Mobile Office*, December: 55–60.

Op. cit. 1994. "On-line sexual advances: how to fend them off." *Glamour,* October, 92 (10): 101.

Op. cit. 1994. "Saving time and money on-line." *Glamour*, September, 92 (9): 129.

Op. cit. 1994. "Use technology to get help online: Online services give you access to managers around the country who have dealt with every imaginable sticky situation." *Executive Female*, July/August, 18.

Op. cit. 1994. "Stalkers in cyberspace: Some strange rangers are hanging out on the electronic frontier. Are they harmless or hazardous to your online safety?" *Online Access*, September: 76–79.

Op. cit. 1993. "Lurkers & flamers: why they do what they do." *Online Access*, June: 48–51.

Dibbell, Julian. 1993. "Rape in cyberspace: a tale of crime and punishment on-line." *Village Voice*, December 21, 1993: 36.

Diwanji, Pavani. 1995. "Corporate supported women's groups." *Communications of the ACM* 38 (1): 36.

Eccles, J.S. 1987. "Gender roles and women's achievement-related decisions." *Psychology of Women Quarterly* 11 (2): 135–171.

Edwards, Paul. 1990. "The Army and the microworld: computers and the politics of gender identity." *Signs* 16 (1): 102–127.

Find/SVP. 1994. "U.S. women and new media." *Interactive Consumers* 1 (2): 1–8.

Frenkel, Karen A. 1990. "Women and computing." *Communications of the ACM* 33 (11): 34–46.

Frissen, Valerie. 1992. "Trapped in electronic cages? Gender and new information technologies in the public and private domain: an overview of research." *Media, Culture and Society* 14 (1992): 31–49.

Herring, Susan. 1993. "Gender and democracy in computer-mediated communication." *Electronic Journal of Communication* 3 (2): Internet.

Herz, J.C. 1994. "Confessions of an Internet junkie." *Playboy* 41(6): 78–84.

Huff, Charles W. and Cooper, Joel. 1987. "Sex bias in educational software: the effect of designers' stereotypes on the software they design." *Journal of Applied Psychology* 17 (6): 519–532.

Inkpen, K., Upitis, R., Klawe, M., Lawry, J., et. al. 1995. "We have never-forgetful flowers in our garden: girls' responses to electronic games." *Journal of Computing, Math & Science Education* (pending publication).

Kantor, Andrew. 1994. "Learning the ropes: a Usenet style guide." *Internet World* (Nov-Dec 94): 24.

Kantrowitz, Barbara with Debra Rosenberg, Patricia King, and Karen Springen. "Men, women and computers: the sexual politics of cyberspace." *Newsweek*, May 16, 1994: 48.

Kantrowitz, Barbara with Joshua Cooper Ramo and Charles Fleming. 1994. "Sex on the info highway." *Newsweek*, March 14, 1994: 62.

Kiesler, Sara, Sproull, Lee, and Eccles, Jacquelynne. 1985. "Pool halls, chips, and war games: women in the culture of computing." *Psychology of Women Quarterly* 9: 451–462.

Kiesler, Sara. 1986. "The hidden messages in computer networks." *Harvard Business Review* (January–February).

Klawe, Maria and Levenson, Nancy. 1995. "Women in computing: where are we now?" *Communications of the ACM* 38 (1): 29–35.

Kramer, Pamela E. and Lehman, Sheila. 1990. "Mismeasuring women: a critique of research on computer ability and avoidance." *Signs* 16 (1): 158–172.

Licklider, J.C.R., Taylor, Robert, Herbert, E. 1960. "The computer as a communication device." *International Science and Technology*.

Markoff, John. 1989. "Computing in America: A masculine mystique." *New York Times* (February 13, 1989): A1, 7 (front page).

Newman, Richard J. 1994. "The high-tech house hunt." *U.S. News & World Report* 116 (14): 70.

Pamintuan, Aimee. 1994. "Can women's mags change cyberspace?" Jupiter Communications' *Interactive Content*, December 1 (8): 1 (5 pages).

Pearl, Amy. 1995. "Women in computing." *Communications of the ACM* 38 (1): 26–28.

Perry, Ruth and Greber, Lisa. 1990. "Women and computers: an introduction." *Signs* 16 (1): 74–101.

Raben, Joseph, and Karl Signell. 1994. "Serving ads." *Internet World* (November-December 94): 72.

Rinzel, Michael. 1995. "Monthly chat scoreboard." *Interactive Content*, January: 12–13.

Salvador, Roberta. 1994. "The emperor's new clothes; The road to utopia?" *Electronic Learning* 13 (8): 32–42.

Span, Paula. 1994. "The on-line mystique: women and computers . . . is there equity in cyberspace?" *The Washington Post Magazine*, February 27, 1994.

Sproull, Lee and Kiesler, Sara. 1991. "Computers, networks and work." *Scientific American* (September 1991): 116–123.

Turkle, Sherry and Papert, Seymour. 1990. "Epistemological pluralism: style and voices within the computer culture." *Signs* 16 (1): 128–157.

van Zoonen, Liesbet. 1992. "Feminist theory and information technology." *Media, Culture and Society* 14: 9–29.

Varley, Pamela. 1991. "What's really happening in Santa Monica?" *Technology Review*.

Wittig, Michelle. 1991. "Electronic city hall." *Whole Earth Review* 71.

# Results of research studies and surveys

Dell Computer Corporation. 1993. Fear of technology is the phobia of the '90s, part of the Techknowledge in America survey.

Jupiter Communications. 1995. The 1995 consumer online services report: 16–18.

Link Resources. 1994. Home media consumer survey.

MCI Telecommunications Corporation. 1994. Cyberphobia in America: a Gallup survey of white-collar workers for MCI Business Markets.

Stuck, M.F. and Ware, M.C. Attitudes toward computers: An exploration of gender differences among undergraduates.

Truong, Hoai-An, et. al. 1993. Gender issues in online communication. BAWIT (Bay Area Women in Telecommunications) paper presented at Third Conference on Computers, Freedom and Privacy: Burlingame, CA.

Ware, M.C. and Stuck, M.F. 1991. Empowering women faculty: a case study of microcomputer use by women faculty and staff. Report, 12th Educational Computing Organization of Ontario Conference and 8th International Conference on Technology and Education: Ontario.

# Books

American Association of Individual Investors Staff. 1994. *The Individual Investor's Guide to Computerized Investing*, 12th edition. Chicago, IL: International Publishing Corporation.

Berkman, Robert I. 1994. *Find It Online!* Blue Ridge Summit, PA: Windcrest/McGraw-Hill, Inc.

Boone, Mary E. 1993. *Leadership and the Computer.* Rocklin, CA: Prima Publishing.

Canter, Laurence A. and Martha S. Siegel. 1994. *How to Make a Fortune on the Information Superhighway.* New York, NY: HarperCollins Publishers, Inc.

Delphi Internet Services Corporation. 1994. *Delphi: the Official Guide.* Cambridge, MA: Delphi Internet Services Corporation.

Dern, Daniel P. 1994. *The Internet Guide for New Users.* New York, NY: McGraw-Hill, Inc.

Dunlop, Charles and Kling, Rob, editors. 1991. *Computerization and Controversy: value conflicts and social choices.* San Diego, CA: Academic Press, Inc.

Ellsworth, Jill H. and Matthew V. Ellsworth. 1994. *The Internet Business Book.* New York, NY: John Wiley & Sons, Inc.

Fox, David. 1995. *Love Bytes: the online dating handbook.* Corte Madera, CA: Waite Group Press.

Gagnon, Eric. 1994. *What's on the Internet?* Fairfax, VA: Internet Media Corporation.

Gilster, Paul. 1994. *Finding It on the Internet: the essential guide to archie, Veronica, gopher, WAIS, WWW, and other search and browsing tools.* New York, NY: John Wiley & Sons, Inc.

Glossbrenner, Alfred. 1995. *Internet 101: A college student's guide.* Blue Ridge Summit, PA: Windcrest/McGraw-Hill, Inc.

Gunn, Angela. 1994. *Plug-n-Play Mosaic.* Indianapolis, IN: Macmillan Publishing/SAMS.

Gunn, Angela, et. al. 1995. *Web Guide: exploring the weird, the wild, and the wonderful on the World Wide Web.* Indianapolis, IN: Macmillan Publishing/SAMS.

Hahn, Harley and Rick Stout. 1994. *The Internet Yellow Pages.* Berkeley, CA: Osborne McGraw-Hill.

Handy, Charles. 1990. *The Age of Unreason.* Boston, MA: Harvard Business School Press.

Hardison, O. B., Jr. 1990. *Disappearing through the Skylight: culture and technology in the 21st century.* New York, NY: Penguin Books USA Inc.

Krebs, Nina Boyd. 1993. *Changing Women, Changing Work.* Aspen, CO: MacMurray & Beck Communications.

Krol, Ed. 1994. *The Whole Internet User's Guide*, special edition. Sebastopol, CA: O'Reilly & Associates, Inc. and SPRY, Inc.

Lambert, Steve and Walt Howe. 1993. *Internet Basics: your online access to the global electronic superhighway.* New York, NY: Random House Electronic Books.

LaQuey, Tracy. 1994. *Internet Companion Plus: a beginner's start-up kit for global networking.* Reading, MA: Addison Wesley.

Levine, John R. and Baroudi, Carol. 1993. *The Internet for Dummies*. San Mateo, CA: IDG Books Worldwide, Inc.

Madara, Edward J. and Abigail Meese, editors. 1990. *The Self-Help Sourcebook*, 3rd edition. Denville, NJ: Saint Clares-Riverside Medical Center.

Magid, Lawrence J. 1994. *Cruising Online: Larry Magid's guide to the new digital highways*. New York, NY: Random House, Inc.

Maloni, Kelly, Ben Greenman and Kristin Miller. 1995. *Net Money: your guide to the personal finance revolution on the information highway*. New York, NY: Michael Wolff & Company and Random House Electronic Publishing.

Maloni, Kelly, Nathaniel Wice, and Ben Greenman. *Net Chat: your guide to the debates, parties and pick-up places on the electronic highway*. New York, NY: Michael Wolff & Company and Random House Electronic Publishing.

Orenstein, Glenn S. and Ruth M. Orenstein. 1994. *CompuServe Companion: finding newspapers and magazines online*. Needham Heights, MA: BiblioData.

Peal, David. 1994. *Access the Internet*. Alameda, CA: SYBEX.

Postman, Neil. 1993. *Technopoly: the surrender of culture to technology*. Vintage Books Edition. New York, NY: Random House, Inc.

Resnick, Rosalind and Dave Taylor. 1994. *The Internet Business Guide*. Indianapolis, IN: Macmillan Publishing/SAMS.

Rheingold, Howard. 1994. *The Virtual Community: homesteading on the electronic frontier*. New York, NY: HarperPerennial.

Rimmer, Steve. 1995. *Planet Internet*. Blue Ridge Summit, PA: Windcrest/McGraw-Hill.

Rutten, Peter, Albert F. Bayers III, and Maloni, Kelly. 1994. *Net Guide: your map to the services, information and entertainment on the electronic highway*. New York, NY: Michael Wolff & Company and Random House Electronic Publishing.

Schepp, Brad and Debra. 1995. *The Telecommuter's Handbook*. New York, NY: McGraw-Hill, Inc.

Sproull, Lee and Sara Kiesler. 1991. *Connections: new ways of working in a networked organization*. Cambridge, MA: MIT Press.

Toffler, Alvin. 1990. *Power Shift*. New York, NY: Bantam Books.

Turkle, Sherry. 1985. *The Second Self: computers and the human spirit*. New York, NY: Touchstone, Simon & Schuster, Inc.

# Videos

Colcott, Glenn. 1994. *The Internet Video*. Osprey Film Productions, Inc.

# Index

# About the author

Judith Broadhurst had written more than 500 articles on online communication, the psychological and social effects of technology, small-business management, and the arts for *Columbia Journalism Review, Executive Female, Glamour, Home Office Computing, Mobile Office, Online Access, Self, Working Woman,* and many other magazines, as well as daily newspapers coast to coast. She's also the editor and publisher of *Freelance Success,* a subscription newsletter for experienced freelance journalists that she publishes online. Judith has taught magazine writing at the New School for Social Research in New York City and has spoken about online networks at many writers' conferences. She currently teaches an online course called "The Basics of Successful Freelance Journalism" and conducts the CyberSavvy Seminars, about useful, real-life ways that women, publishers, and small businesses can use online services. She has received awards from the International Association of Business Communicators, the Public Relations Society of America, Women in Communications, the United Way, and others, and serves on the contracts and online committees of the prestigious American Society of Journalists and Authors (ASJA). She's also a member of the Author's Guild and the Newsletter Publishers Association.

In previous incarnations, Judith spent five years as a flight attendant for Delta Airlines, then several years marketing and fundraising for health and human service agencies and hospitals, followed by similar positions with performing arts organizations. She even booked jazz musicians and produced and promoted festivals and concerts for about six years, but still can't explain what possessed her to get into such a crazy business, rather than just enjoy the music. She has traveled extensively, living in more cities than she cares to remember. She has finally settled, she swears, in the Santa Cruz mountains, south of San Francisco.

Judith is listed in the book *E-Mail Addresses of the Rich and Famous*, much to her surprise. "Rich and famous" is clearly inaccurate in her case, she says, but hopes it's a good omen. To book her for speaking engagements or workshops, please write CyberSavvy Seminars, P.O. Box 1917, Boulder Creek, CA 95006. You can also reach her online at 74774.1740@compuserve.com or broadhurst@aol.com.